KEY GEOGRAPHY

New
Foundations

DAVID WAUGH & TONY BUSHELL

First edition published in 1991 and second edition in 1996 by:
Stanley Thornes (Publishers) Ltd

This edition published in 2001 by:
Nelson Thornes Ltd
Delta Place
27 Bath Road
CHELTENHAM
GL53 7TH
United Kingdom

05 / 10 9 8

A catalogue record for this book is available from the British Library

First edition ISBN 0 7487 1100 7
Second edition ISBN 0 7487 2584 9
Third edition ISBN 0 7487 6041 5

Illustrations by Tim Smith, Kathy Baxendale, Nick Hawken, John Yorke,
Richard Morris, Angela Lumley
Second edition designed and reset by Hilary Norman, third edition designed by
Clare Park
Edited by Katherine James
Photo research by Julia Hanson, Penni Bickle
Printed and bound in China by Midas Printing International Ltd.

The previous page shows St Paul's Cathedral
and the City of London.

Acknowledgements

The authors and publishers are grateful to the following for permission
to reproduce photographs and other material in this book:

Aerofilms Ltd (pp 11, 53C, 56A and B, 96C bottom right); Air Photos
(pp 55C, 56C); Australian Overseas Information Service (pp 79B);
Penni Bickle (pp 42A2, 68A); The Bridgeman Art Library (pp 101D and
E, 102A and B); Channel Tunnel Group Ltd (pp 74,75); Connors/Nigel
Bowles (pp 40A); James Davis Travel Photography (pp 94C); Dundee
University Satellite Receiving Station (pp 24B, 26C); Environmental
Images (pp 42A3); Eye Ubiquitous (pp 42A4 and 5, 68B, 85C, 102C);
Chris Fairclough Colour Library (pp 52A); Forestry Commission (pp
96C top left); Jason Hawkes/Julian Cotton Picture Library (pp 64C);
The Hutchison Library (pp 36C, 44A1, 53C, 99E); Landscape Only (pp
10B); London Aerial Photo Library (pp 64C); Arthur Meredith (pp 68A
middle); National Express (pp 68A right); Panos Pictures (pp 28B,
46A/1, 46A/6; Science Photo Library (pp 104A); Spectrum Colour
Library (pp1); Liz Staves/Collections (pp 25); Still Pictures (pp 43A/1,
44A/2, 46A/2 and 3, 47D, 71C, 82A, 100A and C); Tower Hamlets
Local History Library (pp104A and B); Trip Photography (pp 100B); Dr
A C Waltham (pp 38A, 70A, 95B, 105); Simon Warner (pp 52A and B,
53D, 70A, 94B, 96C top right); York and County Press (pp 42A/6)

Other photos are from the collections of David Waugh and Tony
Bushell

Blockbusters name and grid reproduced by kind permission of Central
Independent Television plc in association with Mark Goodson
Productions and Talbot Television Ltd (pp. 5, 90, 91). The map extract
on p57 is reproduced from the 1992 Ordnance Survey 1:50 000 map
(Landranger 202) and the map extracts on pp. 97, 106 and 109 are
reproduced from the 1994 Ordnance Survey 1:50 000 map of
Cambridge (Landranger 154) with the permission of the Controller of
Her Majesty's Stationery Office © Crown Copyright.

Every effort has been made to contact copyright holders and we
apologise if any have been overlooked.

Contents

What is physical geography?

A

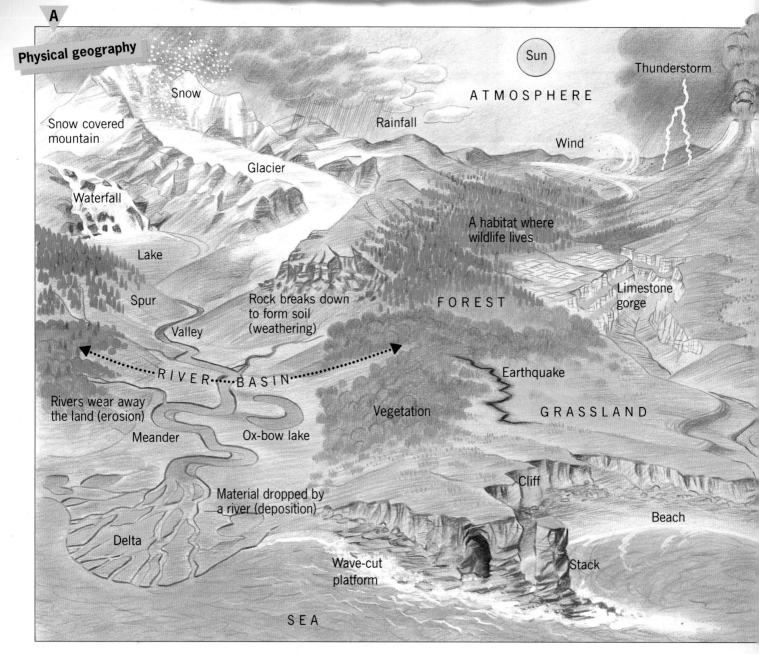

Physical geography

- Sun
- ATMOSPHERE
- Thunderstorm
- Snow
- Rainfall
- Snow covered mountain
- Wind
- Glacier
- Waterfall
- A habitat where wildlife lives
- Lake
- Limestone gorge
- Spur
- Rock breaks down to form soil (weathering)
- FOREST
- Valley
- RIVER BASIN
- Earthquake
- Rivers wear away the land (erosion)
- Vegetation
- GRASSLAND
- Meander
- Ox-bow lake
- Material dropped by a river (deposition)
- Cliff
- Beach
- Delta
- Wave-cut platform
- Stack
- SEA

Geography is the study of the earth's natural features. It is also about people and places and how they affect each other. Geography can help us to understand our world and, hopefully, to make it a better place in which to live.

One of the best ways to learn about geography is to ask questions. You will notice that most pages in this book start with a key question. For example:

- What or where is it?
- What is it like?
- How did it get like this?
- How is it changing?
- What might be the effects of these changes?
- What do I think about them?

There are three main parts to geography. These are physical, human and environmental.

Physical geography is the study of the earth's natural features. It is about the land and the sea and the atmosphere around us.

The **atmosphere** is the air around the earth. Changes in temperature, rainfall and pressure give us our **weather** and **climate**. Climate changes between seasons and from year to year. Different parts of the world have different climates.

Landforms are natural features formed by rivers, the sea, ice and volcanoes. They are continually changing as they are worn away in some places and built up in others.

Most changes in physical geography happen very slowly. Sometimes when sudden changes happen, they cause **hazards** such as storms, floods, drought, volcanic eruptions and earthquakes.

The earth's surface is made up of many different kinds of rock. Where these rocks break up into small pieces, they form soil. Plants grow in this soil and cover most of the earth's land surface.

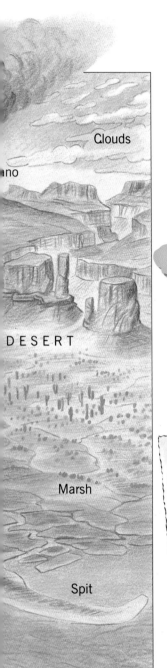

Activity

Make the *Blockbusters Gold Run* by solving the following clues to get a linking route across the puzzle. The letters in the shapes are the start of words you will find in the drawing and written in **bold** on these two pages.

Start at the left-hand side and make your way across the puzzle. When you solve a clue, write down the answer and, if you have a copy of the puzzle, shade in the shape.

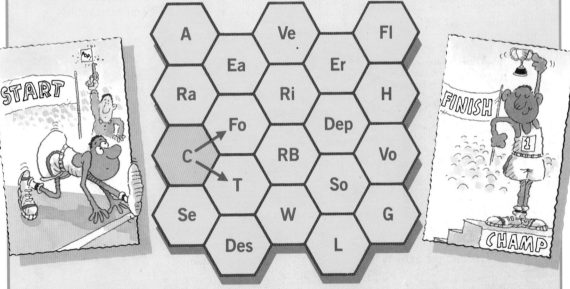

A the air around us
C temperature, moisture and pressure gives us this
Se rivers flow into this
Fo an area covered with trees
Des an area with very few plants
Ri rain collects in these
W how rocks break into smaller pieces
Dep when material is dropped by rivers
L formed by rivers, volcanoes, ice and the sea

H a place where animals and birds live
G a type of vegetation between forest and desert
Fl a physical hazard
Ra the moisture in the air
Ea a physical hazard
T the sun causes this to rise
Ve live plants
RB an area drained by a river
Er the wearing away of the land
So small pieces of rock
Vo a landform and a hazard

What is human geography?

This is the study of where and how people live.

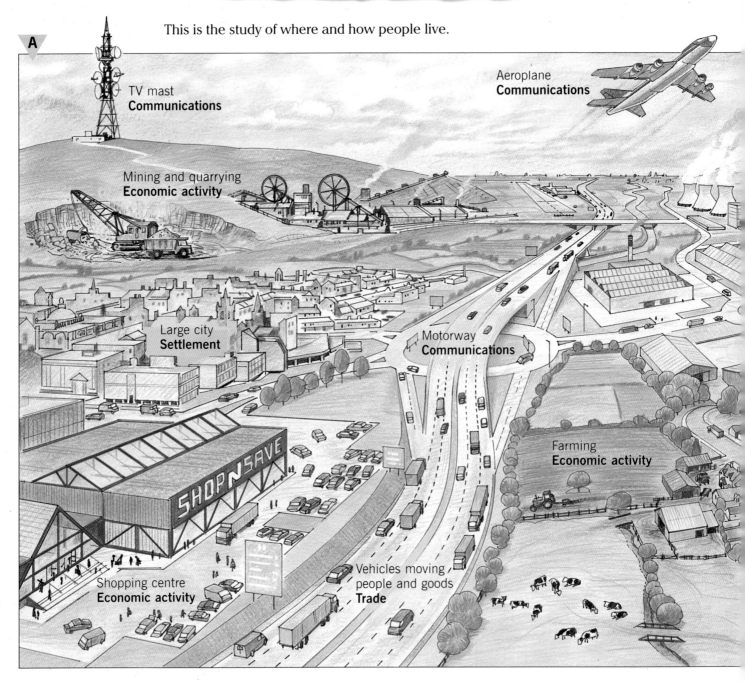

A

TV mast
Communications

Aeroplane
Communications

Mining and quarrying
Economic activity

Large city
Settlement

Motorway
Communications

Farming
Economic activity

SHOP N SAVE

Shopping centre
Economic activity

Vehicles moving
people and goods
Trade

Population geography looks at the spread (distribution) of people over the earth's surface. It tries to explain why some parts of the world have many people living there while other parts have very few. It studies areas where the numbers of people living there are growing rapidly, and looks at the problems that come from this growth. It suggests reasons why people move from one area or country to another (**migration**).

It looks at how such movements lead to people having different customs, religious beliefs and ways of living.

Settlement geography is about where people live. It looks at why settlements grow up in a particular place, and why some remain small in size (villages) while others may grow into very large **urban** centres (cities). It describes problems that go with living in very small places as well as those of very large ones. It looks at how land is used in cities and how this use can change over time.

Communications describes the methods of transport by which people may move about – to work, to school, to the shops and for recreation and holidays. It also includes the movement of goods (**trade**) and information, such as conversations on the telephone and programmes on the television.

Economic geography (economic activity) looks at how people try to earn a living. It is about industry, about jobs and about wealth. It is usually divided into three types. These include farming (a **primary** activity); making things in a factory (a **secondary** activity); or looking after people (a **tertiary** activity or service). It looks at why some activities are only found in certain places and why some parts of the world are richer and more developed than others.

People are very concerned with their **quality of life**. This might be how happy or content they are; the amount of money they have; or how much they like living and working in a particular area. The quality of life may differ greatly both within a country and between countries.

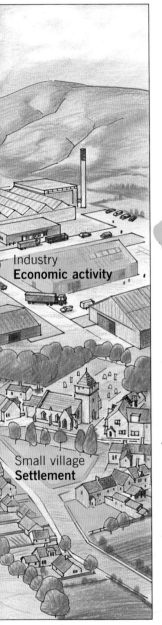

Industry
Economic activity

Small village
Settlement

Activity

On a copy of the *Kriss Kross Puzzle*, solve the puzzle by fitting 11 words and phrases into their correct position in the diagram. The 11 words and phrases are those written in **bold** on these two pages. Two have been done to help you get started.

QUALITY OF LIFE

MIGRATION

7

What is environmental geography?

The **environment** is the combination of the **physical** (natural) environment of climate, landforms, soils and vegetation, and the **human** environment which includes settlements and economic activities. It is the study of the surroundings in which people, plants and animals live.

A

Clean river

Port

Fishing

Sheltered bay protects ships from storms

Areas of scenic value attract tourists

Headlands for walks

Holiday resort with large hotels and amenities

Pier

The environment includes natural **resources** such as coal and iron ore, soils, forests and water. These are used to meet human needs. Some of these resources are **renewable**. This means that they can be used over and over again, such as rainfall. Others are **non-renewable** and can only be used once, such as coal. Sometimes people use these resources to their advantage. For example they use water for drinking purposes, iron ore in industry, and landforms such as islands or lakes for leisure. People often misuse these resources by using them up (minerals), by destroying them (soils, forests) or polluting them (rivers, seas and the air).

Different environments have different qualities and different uses. Each needs to be **protected** and carefully **managed**, like National Parks and the reserves of oil. Many environments have been damaged in the past. Those which have, such as mining areas, rivers and the older parts of some cities, need to be improved.

There is now a growing concern over the **quality of the environment** and how it may be **conserved** while at the same time being made as useful as possible.

B

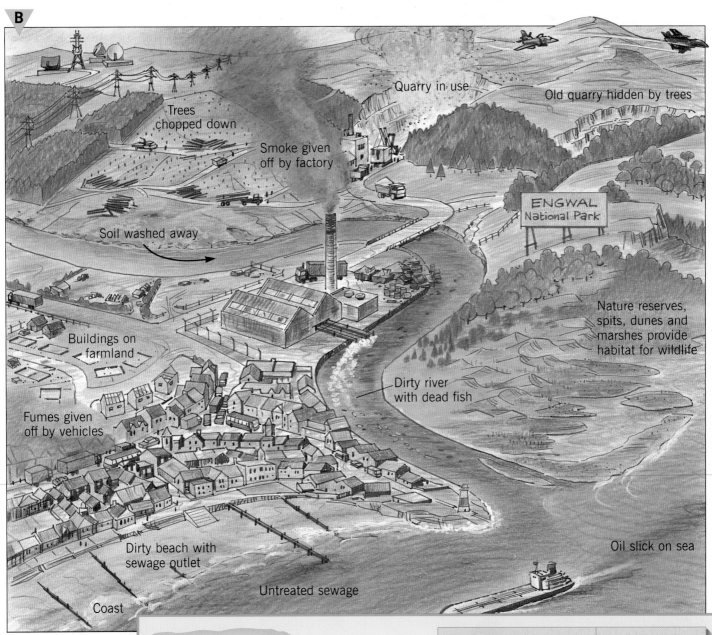

Quarry in use

Old quarry hidden by trees

Trees chopped down

Smoke given off by factory

ENGWAL National Park

Soil washed away

Nature reserves, spits, dunes and marshes provide habitat for wildlife

Buildings on farmland

Dirty river with dead fish

Fumes given off by vehicles

Dirty beach with sewage outlet

Oil slick on sea

Untreated sewage

Coast

Activities

Physical (natural) environment	Human environment
River	Town

1 **a)** Make a copy of the table on the right.
   b) List the features shown in the drawing **A** in the two columns. Two of these features have already been named for you.

2 In what ways has the area shown in drawing **B** been
   a) polluted or destroyed
   b) protected?

How do we study geography?

Geographers need to know about **places**. They should be able to describe where a place is found (located), why it is there (site) and what it might be like to live or work there. Places can include physical features like rivers, mountains and deserts. Places can also be made by humans, e.g. houses, cities, and roads.

Places can vary in **size**. Just as the classroom is a place in a school, so is the school a place in a town, the town a place in a country, and the country a place in the world.

A

Diagram **A** is a plan of a classroom. A classroom is a **place** in a school. If the classroom is neat and tidy (like your bedroom at home!) everything will have its own place. The teacher will have a desk, atlases will be kept on a shelf or in a cupboard, and chalk in a box.

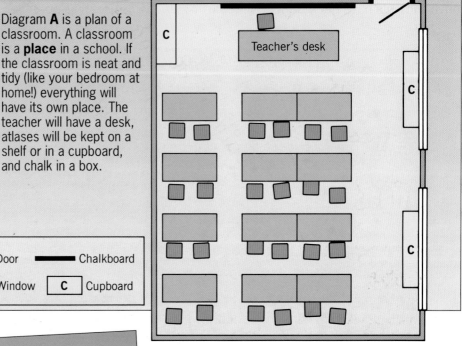

Key			
Chair	• / Door	▬▬ Chalkboard	
Desk	▭ Window	C Cupboard	

In the distance are some low **hills** which are partly covered in **trees**. There is a **village** in the centre of the photo, with a **church** and several **buildings**. The church has a **tower**. A small **road** passes through the village. In front of the village is a **river** which is crossed by a **bridge**. The land around the village consists of **fields** in which **grass** appears to be growing. The fields are separated by **stone walls** and a few **deciduous trees**. The photo was taken on a **sunny** day in **summer** in the **country**.

How do we describe what a place looks like?

Although no two places in the world are exactly the same, they may have similarities. We have to learn how to describe one place so that we can compare it with a second place. We can show how they are similar to or different from each other.

The best way is to use a photo, possibly from a book or a magazine. When writing a description it is important to pick out **key words**. Key words are the important ones to learn and to remember. In the description below photo **B**, the key words have been written in **bold** type so that they are easier to pick out.

We can also describe a place by drawing a labelled (annotated) fieldsketch. Sketch **C** on page 11 is drawn from photo **B**. The labels on the fieldsketch are very similar to the key words in the written description.

C

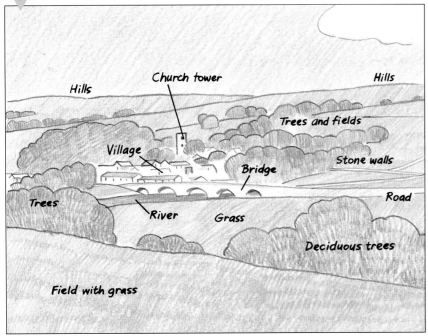

Hills
Church tower
Hills
Trees and fields
Village
Stone walls
Bridge
Road
Trees
River
Grass
Deciduous trees
Field with grass

D

Activities

1 Photo **D** was taken in central London. The key words have been missed out of this description and are listed at the end.

Copy out the description in **E**, putting the key words in the correct places.

2 **a)** Make a copy of the tables below.
  b) Complete the table for photo **B** by adding the key words from fieldsketch **C**.
  c) Complete the table for photo **D** by using the key words listed in question 1.

E

In the distance is a _____ with a _____ and several _____ . Next to the river there is a _____ . In the centre of the photo there are several _____ _____ which appear to be _____ . There are _____ full of _____ . The photo was taken on a _____ day in a big _____ .

bridge   castle   city   offices

roads   sunny   ships   river

tall buildings   traffic

Photo B	
Physical features	Human features

Photo D	
Physical features	Human features

3 **a)** Which photo has more physical features than human features? Why?
  b) Which photo has more human features than physical features? Why?

4 Which of the two places shown in photos **B** and **D** would you rather visit or live in? Try to give reasons for your answers.

E X T R A

Using photo **D** of London, find out the name of the river, the bridge, the castle and the tallest building.

Summary

Geographers study places where people live and those they avoid. Places can be described from photos or by drawing a labelled sketch and underlining key words.

How can we find out where places are?

People often need to know where places are. They need to know this if, for example, they are going shopping or on holiday. Many people, like lorry drivers and ambulance drivers, need to know where places are to do their job. On television we are always hearing about different places, on the News and in other programmes.

Geographers use maps to find out where places are and what they are like. An **atlas** is a book that has maps showing places all around the world, and it is easy to use. The most accurate way to show the whole world is on a **globe**. This is because a globe, being round, shows the actual shape of the earth.

To help us find places, imaginary lines called **latitude** and **longitude** are drawn onto the globe. These are shown in diagram **A**.

A

Latitude

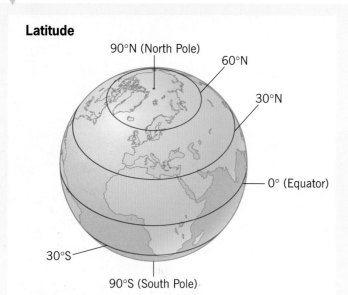

Lines of latitude are imaginary lines going around the earth from east to west. They are measured in degrees north or south of the **Equator**. Latitude 0° is called the Equator.

Longitude

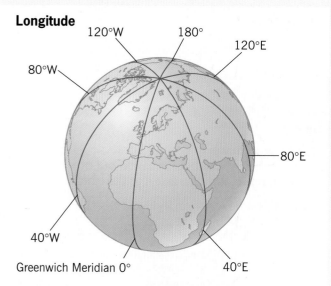

Lines of longitude are imaginary lines going from the North Pole to the South Pole. They are measured from the **Greenwich Meridian**, which passes through London. The Greenwich Meridian is 0°.

It is impossible to draw the earth accurately on a piece of paper. Parts of it will always be either the wrong size or the wrong shape. This is because the earth is round, and a piece of paper is flat. Imagine peeling the skin off an orange and trying to lay it out flat. It cannot be done, because the peel will split and some parts will be pushed out of shape.

One way of drawing the globe as a flat map is shown in map **B**. Some places have been stretched, and others squashed to make them fit.

B

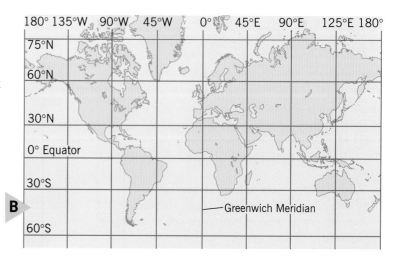

Using an atlas

The **contents** page at the front of the atlas shows on which page each map can be found. The **index** at the back of the book shows exactly where a particular place may be found. The index gives the latitude and longitude of that place to help you find it more easily. Diagram **C** shows you how to use the index of an atlas.

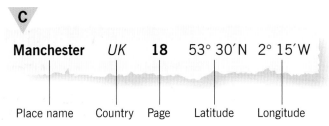

C

| **Manchester** | *UK* | **18** | 53° 30′N | 2° 15′W |

Place name   Country   Page   Latitude   Longitude

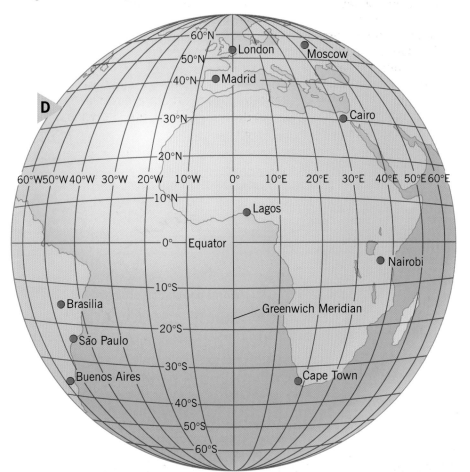

D

✔ Remember

✔ Lines of latitude go across the map.
✔ Lines of longitude go up and down the map.
✔ Latitude is always given first.
✔ Latitude and longitude are measured in degrees (°). Each degree is divided into 60 smaller parts called minutes (′).

Activities

1 Look at map **D** above. Name the city at each of the following:
a) 30°N 31°E   b) 34°S 18°E
c) 40°N 4°W   d) 24°S 47°W.

2 Use map **D** to give the latitude and longitude for each of these cities:
a) London   b) Lagos   c) Moscow
d) Buenos Aires   e) Nairobi   f) Brasilia.

3 Find each of the cities below in the index of your atlas. For each one give the country, page number and latitude and longitude.

| New York | Tokyo | Sydney | Calcutta |

Summary

Maps are useful to people. They help us to find out where places are and what they are like. An atlas shows many places around the world. These places may easily be found using latitude and longitude.

How can we use graphs in geography?

Graphs are diagrams that show information in a clear and simple way. They can be used to describe a situation and show how one thing is related or linked to another.

Graphs are drawn using facts and figures which are called **data**. We can obtain data either by collecting information from fieldwork or by looking it up in a book. Information that we collect ourselves is called **primary data**. Information from other sources is called **secondary data**.

Graphs can either be drawn by hand or on a computer using a spreadsheet program.

✔ Remember

When you draw a graph, it should have:
✔ a title to say what it is showing
✔ labels along the bottom and up the side to explain what they are showing
✔ figures that are plotted very accurately.

A

Bar graphs

Rainfall graph for London

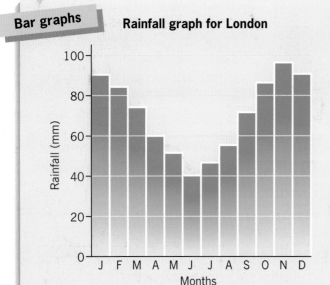

- A bar graph is made up of several bars or columns.
- The bars can be drawn either horizontally across the page or vertically up and down the page.
- Bar graphs are used to compare different things or quantities.
- The graph above compares the amount of rain in each month of a year. It shows which parts of the year are wettest.

B

Line graphs

World population change

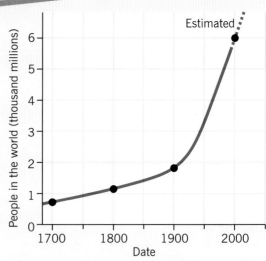

- Line graphs show information as a series of points that are joined up to form a line.
- Line graphs show changes or trends over a period of time. They can also help predict or forecast future changes.
- The graph above shows that population grew slowly between 1700 and 1800, then very rapidly between 1900 and 2000. The graph also suggests what might happen in the future.

Activities

1 Look at graph **A**.
   a) Name the four months with least rainfall.
   b) How much rainfall was there in November?

2 Look at graph **B**.
   a) What was the population in 1700?
   b) What does the graph suggest might happen to population in the future?

3 Look at graph **C**.
   a) In which season are fewest holidays taken?
   b) What percentage of holidays are taken in summer?

4 Look at graph **D**.
   a) How much rainfall gave a river depth of 2 metres?
   b) Which point, A, B or C, shows there to be little rainfall and a small amount of water in the river?

5 Which type of graph would be best for:
   a) showing the change in the cost of petrol
   b) comparing the size of UK cities
   c) showing how the land use of an area is divided up?

Summary

Graphs are diagrams used to show data clearly. The four types of graph are the bar graph, line graph, pie chart and scatter graph.

C

Pie graphs

People taking holidays in the UK

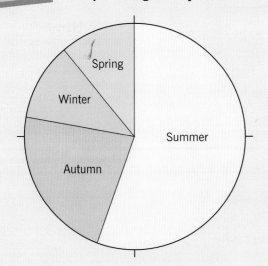

- A pie graph is drawn as a circle which is then divided into several pieces or sectors.

- The whole circle is always equal to 100%.

- Pie graphs show proportions and help us to see how something is divided up.

- The graph above shows that more than half of the people in the UK take their holidays in the summer months.

D

Scatter graphs

Rainfall and river depth

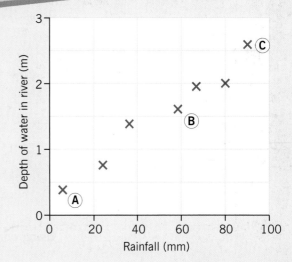

- A scatter graph has data plotted as a number of dots or crosses.

- Scatter graphs are used to see if information about two different things is related or linked.

- The graph above shows the link between rainfall and the amount of water in a river. As rainfall increases, so the amount of water in the river increases.

How can we use computers in geography?

Computers are useful to geographers and can be used in many different ways. At a simple level they can be used to write text, draw diagrams and find out information. At more advanced levels they may be used for such things as forecasting the weather, predicting where and when floods might occur, and researching how earthquakes and volcanic eruptions happen. Computers may also be used to communicate information.

Some ways that computers, or **Information and Communication Technology (ICT)**, can be used in geography are shown below.

A

A **word processor** can be used to write notes, reports or essays about a topic that you are studying. The text can easily be changed, printed out or saved for later use.

The **internet** is useful for geographical research. It can provide up-to-date statistics and information on almost every geographical topic.

A **spreadsheet** program can be used to draw graphs and diagrams. These show statistical information clearly and are most useful when completing a geographical investigation.

CD-ROMs are like electronic books. They contain large amounts of information about particular topics. Most are interactive and some can be used to simulate, or copy, geographical situations.

Using computers in geography

Desktop publishing (DTP) can be used to present information in a clear and attractive way. It is best used when good design is needed and impact is important.

E-mail is electronic mail. It can be used to communicate with other users quickly and easily. E-mail is excellent for exchanging information and sharing views about a topic or an issue in geography.

The internet is very useful to geographers. It provides almost instant access to people all over the world, and gives up-to-the-minute information on just about every topic and issue in geography.

Much of the information on the internet is updated daily, and it can come in a variety of forms. These include text, statistics, videos, photographs, satellite images, weather charts, diagrams, newspaper reports, eye-witness accounts and even government records.

The internet also allows teachers, pupils and schools from all over the world to communicate with each other, sharing information, ideas and experiences.

Nelson Thornes, the publisher of this book, has developed an internet site especially for Key Geography. It provides resources and information that are directly linked to books in the series. The site also gives website addresses that you might find useful in your studies.

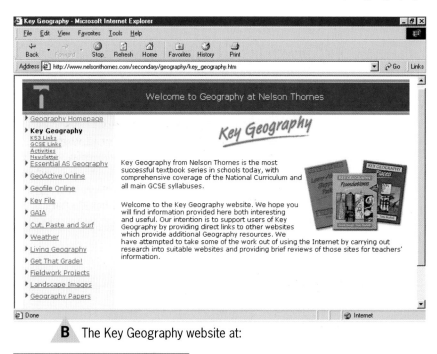

B The Key Geography website at:

www.nelsonthornes.com

then click

secondary/geography/key_geography.htm

Activities

1 **a)** Make a larger copy of table **C**. You will need three lines for each use or application.
   b) Complete the table by adding examples from the list **D** below.
   c) Add an example of your own to each use.

D
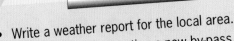

- Write a weather report for the local area.
- Draw a poster supporting a new by-pass.
- Exchange fieldwork data with a colleague.
- Find weather statistics for world cities.
- Make brief notes on a flood disaster.
- Show monthly temperatures for a year.
- Use a program to simulate settlement growth.
- Compare the size of UK cities.
- Produce a front cover for an enquiry.
- Collect information on a flood disaster.
- Send geography homework details to a friend.
- Use an electronic atlas to find places in the world.

C

Computer use or application	Examples
Word processor	
Graphs or diagrams	
Desktop publishing	
Internet	
E-mail	
CD-ROM	

2 **a)** What does ICT stand for?
   b) Why do you think it is useful to use ICT in geography?

3 **a)** Log onto the internet and visit the Key Geography website at the address given above.
   b) Make a list of the pages and features that may be useful to you in your geography studies.

Summary Computers are important in geography. They give access to information sources, and help in presenting and communicating geographical data.

2 Weather and climate

How might you observe and record the weather?

Weather can be described as the condition of the air around us over a short period of time. It is about being hot or cold, wet or dry, windy or calm, cloudy or sunny.

Meteorology is a study of the weather. One of the important tasks of meteorologists is to measure and record all the features of the weather every day. Many expensive and complicated instruments are needed to record weather accurately but you can get a good picture of what conditions are like by **observing** (looking around) and using simple equipment.

A

Temperature
This is a measure of how hot or cold it is. You can do this by looking at the clothes that people are wearing. Thermometers are used to measure temperature accurately.

Very cold

Cold

Mild

Warm

Hot

B

Precipitation
Water in the air falls to the ground in one of several forms. Four of these are rain, snow, sleet and hail.

0°C / 0°C / Below 0°C / Above 0°C

C

Wind speed
This tells us how strong the wind is. We can get a good idea of this by looking at smoke and the trees. The Beaufort scale is used to measure wind strength.

0 Calm	2 Light breeze	4 Moderate breeze	6 Strong breeze	8 Fresh gale

Smoke rises vertically | Wind felt on face; leaves rustle | Dust and paper lifted; small branches move | Large branches in motion | Twigs break off trees

D

Cloud type
Cloud comes in many shapes, sizes and heights. Cumulonimbus, cumulus, stratus and cirrus are the most common types.

Cumulonimbus

Cumulus

Stratus

Cirrus

18

E

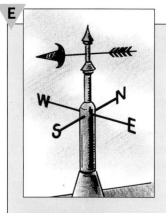

Wind direction
This is the direction **from** which the wind blows. It is measured by a wind vane.

F

○ Clear sky

◔ 2/8

◑ 4/8

◕ 6/8

● 8/8

(total cloud cover)

Cloud cover
This is the amount of the sky covered by cloud. It is measured in eighths.

G

Visibility
This is the distance that can be seen. It is measured in metres.

H

General weather
This describes the weather in words. Words like rain, snow, showers, fog, mist, thunder, cloudy, fair or sunny are used. Light or heavy can be added to precipitation.

Activities

1 What is weather?

2 **a)** Make a copy of the star diagram **I** on the right.
 b) Write the name of the weather feature next to each sketch.

3 Describe how each of the following is measured:

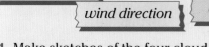

temperature

wind strength

wind direction

cloud cover

4 Make sketches of the four cloud types in **D**. Under each sketch write a cloud description from the following list.
 • Low grey shapeless cloud that forms in layers.
 • High clouds that are wispy, light and featherlike.
 • Dome shaped clouds with dark flat bases.
 • Huge, towering clouds that often give showers.

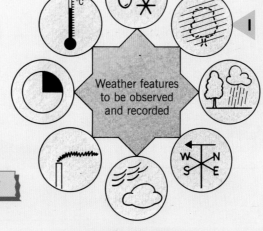

Weather features to be observed and recorded

I

5 Look at the table **J** below which shows what the weather was like on a summer day in Wales.
 a) Copy out the table headings.
 b) Make your own recording of today's weather. Use the information on these two pages to help you.

E X T R A S

1 Keep a record of the weather for a week. Do this at the same time each day. Record your readings in a table.

2 See if you can spot any link between the wind direction and other features of the weather.

J

Day	Temperature	Precipitation	Wind speed	Wind direction	Cloud amount	Cloud type	Weather
Sunday July 15	Warm	Rain showers	Force 2	Westerly	4/8	Cumulus	Mainly sunny with some rain
Monday July 16							

Summary Weather is the day to day condition of the atmosphere. A simple record of the weather may be made by careful observation of what is going on around us.

How can local features affect temperature and wind?

On a fine summer's day are some of the classrooms in your school hotter than others? When the sun shines or a cold wind blows, is one side of your classroom warmer or colder than the other? On a hot sunny day can you notice a difference in temperature between a dark, tarmac playground and a grassy area like the school field? Are there some sheltered places around your school where you can get out of the wind?

Look at cartoon **A** which shows how different the conditions can be on two sides of a hedge.

Each particular place or site tends to develop its own special climate conditions. When the climate in a small area is different to the general surroundings it is called a **microclimate**. Some of the causes of microclimates are given below.

A

B

Shelter

Trees, hedges, walls, buildings and even hills can provide shelter from the wind. Wind speed may be reduced and its direction changed. Places sheltered from cold winds will be warmer.

Physical features

Trees provide shade and shelter and are usually cooler than surrounding areas. Water areas such as lakes and seas have a cooling effect and may also produce light winds. Hill tops are usually cool and windy.

Surface

The colour of the ground surface affects warming. Dark surfaces such as tarmac and soil will become warmer than light surfaces such as grass.

Buildings

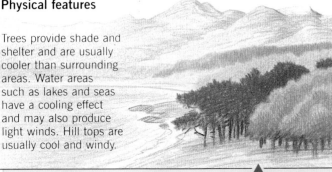

Buildings give off heat that has been stored from the sun during the day or which leaks from their heating systems. Temperatures near buildings may be 2°C or 3°C higher.

Buildings break up the wind and can reduce wind speeds by up to a third. Sometimes the wind can increase speed as it rushes round buildings.

Aspect

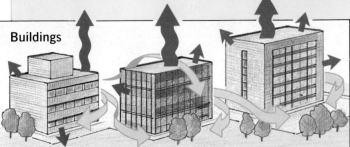

The direction in which a place is facing is called its aspect. Places facing the sun will be warmer than those in shadow.

In Britain the sun rises in the east and moves through the south before it sets in the west. South facing places get most of the sun and are usually the warmest.

Sun at midday

Cool around edge of lake

Main wind direction

Cool in trees with less wind

Cool and windy in shade and facing wind

Grassy play area sheltered from wind

Cooler classrooms due to shade and effect of wind

Hotter classrooms on sunny side of school

Play area warmed by dark tarmac surface

Some warmth from building

C A school's microclimate

Activities

1 Describe a place at your school which is
   a) often sunny
   b) usually in the shade
   c) sheltered on a windy day.

2 Copy and complete diagram **D** by filling in the bubbles with the following words or statements.
   ● Climate conditions of a small area
   ● Physical features
   ● Dark surfaces warm up most
   ● Reduces the effect of wind
   ● Buildings
   ● Aspect

D Local weather conditions

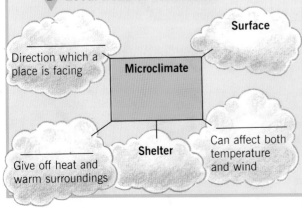

Direction which a place is facing

Surface

Microclimate

Can affect both temperature and wind

Shelter

Give off heat and warm surroundings

3 From photo **C** give **eight** features of the school's microclimate. List your answers under the headings:

   Aspect   Shelter   Others

E X T R A

Microclimate Enquiry

1 **Aim** – to find out what effect aspect has on temperature

2 **Equipment** – thermometer

3 **Method:**
   a) Take several temperature readings on the north and south facing sides of the school. Make a note of the weather each time (e.g. sunny, cloudy, windy).
   b) Make a copy of the table below and display your results.
   c) Describe your findings.
   d) Suggest reasons for your findings.

Time	North facing	South facing	Weather conditions
Average			

Summary

Site conditions such as aspect, shelter, physical features and other factors can influence temperature, local wind speed and direction.

What is Britain's weather?

Weather is what happens in the atmosphere day by day but **climate** is different. It is the weather taken on average over many years. Climate is about warm dry summers, cool wet winters or, as at the North and South Poles, being cold all year. In Britain the weather is always a popular topic of conversation, probably because it is always changing or it's never quite what we want it to be. Changes also occur in the climate. It can change from time to time (seasonal) or it may be different from place to place.

A

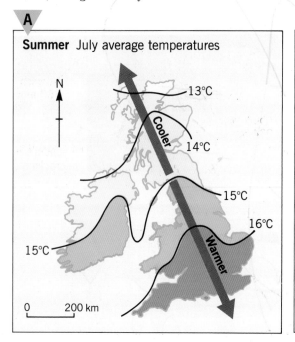

Summer July average temperatures

N

13°C
Cooler
14°C
15°C
16°C
15°C
Warmer

0   200 km

B

Winter January average temperatures

N

5°C
4°C
6°C
Colder
Milder
4°C
5°C
6°C
7°C

0   200 km

The average monthly temperatures for summer and winter are shown on maps **A** and **B**.

If you look closely you should see three main differences.

1 As expected, temperatures are higher in the summer than in winter.
2 Temperatures at any one time are not the same all over Britain.
3 The pattern of temperature is different in the two seasons.

Map **C** shows three important reasons for these variations in weather and climate. Another two are:

1 **Wind direction** – where the air has come from. A north wind will be cold, a west wind will be moist.

2 **Distance from the sea** – the sea keeps coastal places warm in winter but may cool them in summer. Places far inland will have warmer summers and cooler winters.

C

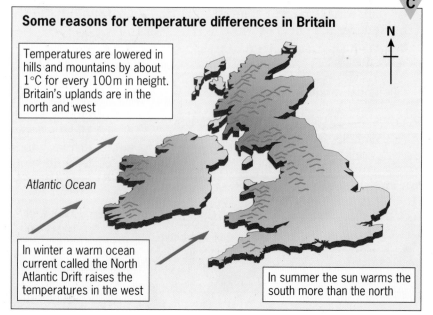

Some reasons for temperature differences in Britain

N

Temperatures are lowered in hills and mountains by about 1°C for every 100m in height. Britain's uplands are in the north and west

Atlantic Ocean

In winter a warm ocean current called the North Atlantic Drift raises the temperatures in the west

In summer the sun warms the south more than the north

Rainfall

In Britain we can expect rain at any time of the year. Although winter is wetter than the summer, seasonal differences in rainfall are very small.

As map **D** shows, however, the amount of rainfall varies considerably from place to place and the greatest differences are between the east and the west.

D

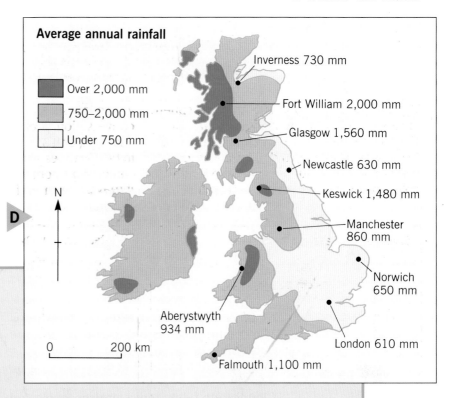

Average annual rainfall

Over 2,000 mm
750–2,000 mm
Under 750 mm

N

Inverness 730 mm
Fort William 2,000 mm
Glasgow 1,560 mm
Newcastle 630 mm
Keswick 1,480 mm
Manchester 860 mm
Norwich 650 mm
Aberystwyth 934 mm
London 610 mm
Falmouth 1,100 mm

0   200 km

Activities

1 What is the difference between weather and climate?

2 **a)** Write out and complete the following sentence to describe summer temperatures in Britain.

> Summers in Britain are _____ than winter. The warmest weather is in the _____ and temperatures get lower (decrease) towards the _____ .

b) Write a similar sentence to describe winter temperatures.

3 Why are there temperature differences in Britain? Think of **three** reasons and write them in your work book.

4 **a)** List the **three** wettest and the **three** driest towns from map **D**. Give your answers in order with the wettest first.
   b) With the help of a simple diagram, describe the difference in rainfall from east to west. Give actual figures in your answer.

5 **a)** Make a large copy of map **E**.
   b) Match the following climate descriptions to areas (A), (B), (C) and (D) and write them on your map. (A) has been done on the map to help you.
   ● Warm summers, cold winters, dry
   ● Mild summers, mild winters, wet
   ● Warm summers, mild winters, quite wet
   ● Mild summers, cold winters, dry

c) Suggest reasons for the climate of area (A).
d) Mark where you live on your copy of map **E**.
e) Describe the climate there and suggest reasons for it.

E

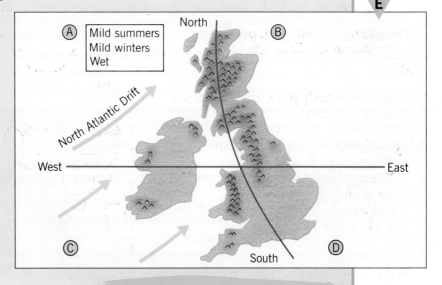

(A) Mild summers / Mild winters / Wet

North
West
East
South

(B)
(C)
(D)

North Atlantic Drift

Summary

Britain's climate varies from place to place and from season to season. Heating from the sun, ocean currents, and the height of the land are some of the reasons for these variations.

How does it rain?

The Atacama Desert in South America has had no rain for over 400 years yet parts of the Amazon rain forest, also in South America, have rain on more than 330 days each year. Seathwaite in the Lake District, the wettest place in England, has on average 3,340 mm of rain per year, whilst Newcastle, only 130 km away, may expect just 630 mm.

What are the reasons for this, what causes rain and why are some places wetter than others?

Clouds are made up of extremely tiny drops of moisture called **cloud droplets**. They are only visible because there are billions of them crowded together in a cloud.

Clouds form when moist air rises, cools and changes into cloud droplets. This is **condensation**. A cloud gives rain after these tiny cloud droplets grow thousands of times larger into raindrops which then fall to the ground.

Look at diagram **A**. It shows how rain is formed. The process is always the same: air rises, cools, condenses and precipitates.

Air can be forced to rise in three different ways. This gives the three main types of

rainfall: **relief, convectional**, and **frontal**. These are shown in diagrams **B, C** and **D**.

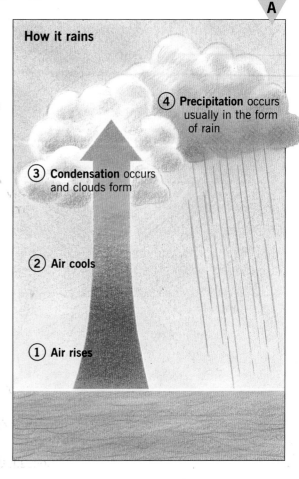

A

How it rains

④ **Precipitation** occurs usually in the form of rain

③ **Condensation** occurs and clouds form

② **Air cools**

① **Air rises**

B

Relief rainfall in the north of England

Relief rainfall occurs when moist air is forced to rise over mountains. As it rises it cools and the rainmaking process shown in diagram **A** comes into operation.

Relief rainfall is quite common in Britain especially in the west where most of the high land is located.

Cloud and rain

Air rises over Lake District mountains

Air descends and warms

Rain stops

North Sea

Warm moist air

• Seathwaite

Irish Sea

Newcastle

Lake District

West

East

C

Convectional rainfall

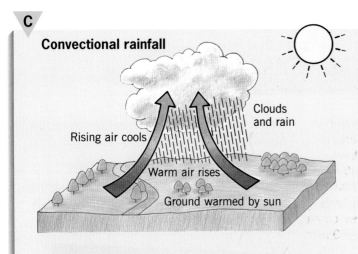

Clouds and rain

Rising air cools

Warm air rises

Ground warmed by sun

When the ground surface is heated by the sun, the air above it is warmed up. This air rises and as it cools down clouds form and rain follows. The showery weather and thunderstorms of a British summer are this type of rainfall.

D

Frontal rainfall

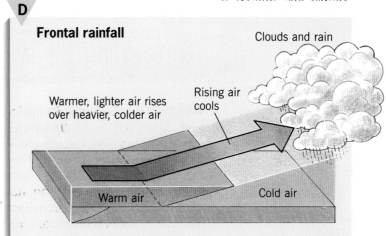

Clouds and rain

Warmer, lighter air rises over heavier, colder air

Rising air cools

Warm air

Cold air

When a mass of warm air meets air at a lower temperature, it rises up and over the colder, heavier air. Once it is made to rise, cloud and rain will follow due to the process shown in diagram **A**.

The place where warm air and cold air meet is called a **front**. Frontal rainfall is very common in Britain throughout the year and especially in winter.

Activities

1 Match the beginnings of these labels to their correct endings.

Clouds are	rain, snow and other forms of moisture in the sky.
Precipitation is	when water vapour changes to water.
Condensation happens	made up of tiny drops of moisture called cloud droplets.

2 With the help of a labelled diagram, describe how it rains.

3 a) Make larger copies of the three diagrams below.
   b) For each diagram explain how it rains by adding labels at points ①, ②, ③ and ④.
   c) Add colour to make your diagrams clearer.
   d) Underneath each of your diagrams give a brief reason for the air rising.
   e) Give each diagram a title.

4 Explain why Seathwaite is wetter than Newcastle. Use diagram **B** to help you.

Summary

Rain is caused by moist air rising and cooling. The three types of rainfall produced in this way are relief, convectional and frontal.

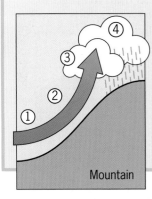

Mountain

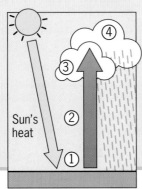

Sun's heat

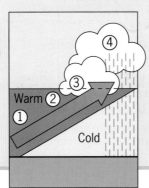

Warm

Cold

Forecasting the weather — anticyclones

Weather has an important effect on our lives. Every day in the newspapers and every evening after the television news there is a **weather forecast**. Forecasts can tell us in advance what the weather will be. For many of us they are of passing interest but for some people such as farmers, fishermen, aircraft pilots and builders the forecasts are very important because the weather affects their work and even their safety.

Map **A** is a typical newspaper weather map. Notice how easy it is to read the weather using the picture symbols.

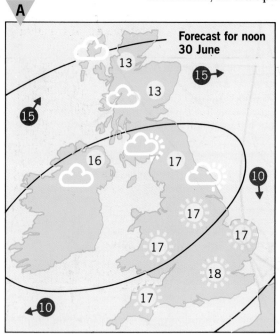

A

Forecast for noon
30 June

13
15→
13
←15
16
17
10↓
17
17
17
18
17
←10

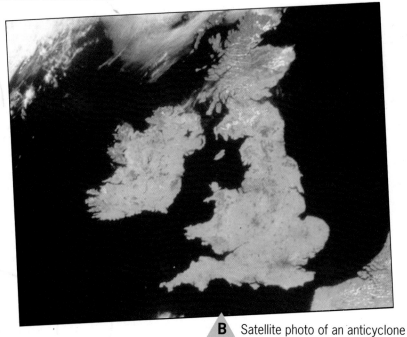

B Satellite photo of an anticyclone

How do weather forecasters know what the weather will be like tomorrow? How can they tell if it will be wet or dry, or hot or cold?

Forecasting is very complicated and lots of information and advanced computers are needed to make good forecasts. In recent times, satellites have become particularly useful because they can see weather systems many kilometres away.

Photo **B** has been taken from a satellite. It shows Britain with very little cloud overhead and clearly enjoying a fine sunny day. Photos like these are taken every few hours and by looking back over several of them, the movements of the weather systems can be worked out, and forecasts made.

The weather system in photo **B** is an **anticyclone**. It occurs because of changes in the air pressure. The weight of air pressing down on us from above is called pressure. This pressure varies from place to place and results in the development of pressure systems. Areas with above average pressure (high pressure) are called anticyclones and usually give good weather. Areas with less than average pressure (low pressure) are called depressions and usually give poor weather.

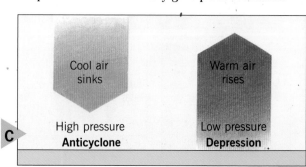

Cool air sinks

Warm air rises

High pressure
Anticyclone

Low pressure
Depression

C

D

Features of an anticyclone

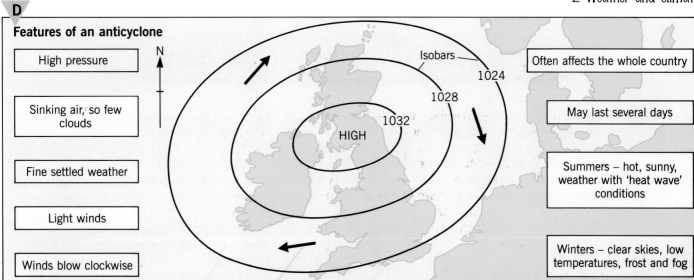

| High pressure |
| Sinking air, so few clouds |
| Fine settled weather |
| Light winds |
| Winds blow clockwise |

Isobars
1024
1028
1032
HIGH
N

| Often affects the whole country |
| May last several days |
| Summers – hot, sunny, weather with 'heat wave' conditions |
| Winters – clear skies, low temperatures, frost and fog |

Activities

1 From map **A**, give the weather that is forecast for the place where you live.

2 **a)** When do you think it would be useful for you to know the next day's weather?
   b) Make a list of people who need the weather forecast. For each person explain why they need to know about the weather.

3 How do satellites help in forecasting weather conditions?

4 **a)** Make a sketch of an anticyclone like the one in diagram **D** above.
   b) Next to your sketch, write out the paragraph below and fill in the blank spaces with the following words:

• LONG   • LARGE   • HIGH   • COOL

Anticyclones are areas of _____ pressure which form when _____ air sinks. They usually cover _____ areas and give _____ periods of fine settled weather.

5 Copy and fill in table **F** to show the weather features of an anticyclone.

E

Weather in a winter anticyclone

F

	Summer	Winter
Temperatures		
Cloud cover		
Wind speed		
Wind direction		
Rain		
Other features		

Summary

Knowing what the weather will be like can be useful to us. Anticyclones can bring good weather and may be forecast with the help of satellites.

E X T R A S

1 Use photo **E** to describe the weather of an anticyclone in winter.

2 Study map **A** on page 26 giving the newspaper weather forecast. Write a weather forecast to be read out on the radio for the same day. Your forecast should be about 100 to 150 words in length.

Forecasting the weather — depressions

All too often we seem to hear the weather forecast begin with 'Today will be cloudy, and rain already in the west will spread eastwards to cover all areas by late afternoon . . .'. The reason for this is that for much of the year Britain is affected by low pressure.

As diagram **A** shows, at times of low pressure the air is usually rising. As it rises it cools, condenses and clouds form. Low pressure areas are called **depressions**. Depressions are the most important weather systems affecting Britain and they bring with them clouds and rain.

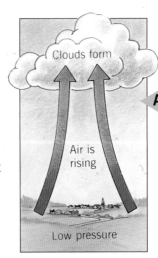

Clouds form

Air is rising

Low pressure

A

Depressions develop where warm air meets cold air. The boundary of the two different air types is called a **front**. Along a front there will be cloud and usually rain. Diagram **B** shows the features of a depression. The isobars are lines that join up areas of equal pressure and they help us to see the shape of the depression.

B

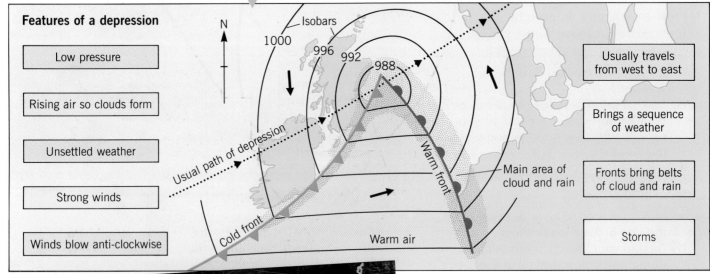

Features of a depression

N

Isobars
1000
996 992
988

Low pressure

Rising air so clouds form

Unsettled weather

Strong winds

Winds blow anti-clockwise

Usual path of depression

Cold front

Warm front

Warm air

Usually travels from west to east

Brings a sequence of weather

Main area of cloud and rain

Fronts bring belts of cloud and rain

Storms

C Satellite photo of a depression

Depressions are huge areas of low pressure measuring many hundreds of kilometres across. They show up very clearly on satellite photographs as great swirls of cloud that look like gigantic catherine wheel fireworks. The fronts are easily recognised as areas of thick white cloud arranged in an upside down 'V' shape. The centre of the depression is normally just above or a little behind the point of the 'V'.

Look at photo **C** which shows a depression approaching Britain. Can you work out which areas are the fronts and where the centre of the depression might be? With help from diagram **B** can you work out which is the area of warm air? What sort of weather does that area seem to have?

Depressions usually form over the Atlantic Ocean and move across Britain from west to east. With help from satellite photographs, weather forecasters can work out the direction they are travelling and how fast they are moving. From this information they can produce quite accurate weather forecasts. Diagram **D** shows how the weather changes as a depression passes over Britain. Notice the changes in the weather that occur in the area where you live.

D A depression passing over Britain

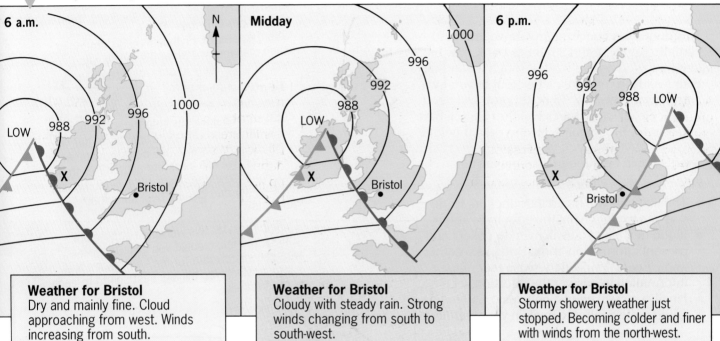

6 a.m.

Weather for Bristol
Dry and mainly fine. Cloud approaching from west. Winds increasing from south.

Midday

Weather for Bristol
Cloudy with steady rain. Strong winds changing from south to south-west.

6 p.m.

Weather for Bristol
Stormy showery weather just stopped. Becoming colder and finer with winds from the north-west.

General features	Weather

Activities

1 The words below have been jumbled up. Unscramble the words and fill in the blank spaces in the following paragraph.

NIRA   SATE   DULOC   OWL   TEWS   SERIS

Depressions are areas of _____ pressure which form when air _____ . They usually move across Britain from _____ to _____ and bring most of our _____ and _____ .

2 With the help of a labelled diagram, explain why depressions bring cloud and rain.

3 Make a labelled sketch of a depression like the one shown in diagram **B**. Underneath your sketch make a copy of the table at the top of the next column.

Complete the table to show the main features of a depression.

4 From diagram **D**:
a) Describe the weather at place **X** for 6 a.m., 12 midday and 6 p.m.
b) Explain why the weather has changed.
c) At what time will the warm front be over the place where you live?
d) Describe the weather you may get at that time.

EXTRA

a) Trace the outline of Britain from photo **C**.
b) Mark and label the following:
■ warm front
■ cold front
■ warm air
■ centre of the depression.
c) Shade the area of cloud and rain along the fronts.
d) Describe the weather over Britain.

Summary

Depressions are the most common weather system affecting Britain. They are low pressure areas and bring stormy winds, cloud and rain.

The weather enquiry

World Wide Leisure Corporation

174 Aspen Boulevard,
Denver,
Colorado 96543, USA

Tel./fax (303)569-8809

Dear Sir/Madam

I am the Personnel Manager for a large American company. We are planning to open four offices in Britain. These offices will be at Oban, Aviemore, Plymouth, and Cambridge. Each manager will bring the family with them, and they are likely to stay for three years.

Like many Americans, each family is keen on leisure and on doing things out of doors. Of course these activities in turn depend upon the weather and climate. Each family has different interests – as will be listed later in this letter. We are therefore allowing each of them to choose the office in the region where the weather and climate best suits their interests.

To help them do this we would appreciate your help. Please could you give us the weather and climate for the four places and suggest which you think is best suited to each of the families.

Yours sincerely

John F. Gates
Personnel Manager

This enquiry is concerned with weather and climate. Pages 22 and 23 of this book may be helpful to you as you work through the enquiry. Your task is to reply to a letter sent to you by a company in America. To do this you will need to look closely at Britain's weather and climate and answer the enquiry question given below.

There should be three main parts to your enquiry.

- The first part will be an introduction. Here you can explain what the enquiry is about.
- In the next part you will need to collect and present information about Britain's weather and climate. You can then use that information to answer the question set.
- Finally you will need a conclusion. Here you could write a letter to explain your findings.

What are the differences in weather and climate across Britain?

1 **Introduction – what is the enquiry about?**
You will need to use maps and writing here. Star diagrams or lists might also help.

a) First look carefully at the enquiry question above and say what you are going to try to find out.
- Give the meanings of **weather** and **climate**. Pages 18, 22 and the Glossary will help you.
- Describe how Britain's weather and climate can be roughly divided into four regions.
- Explain where these regions are, and then describe the different conditions in them.

b) Describe briefly what the letter has asked you to do. Show on a map where the four places are located. List the particular features of weather and climate that you will look at.

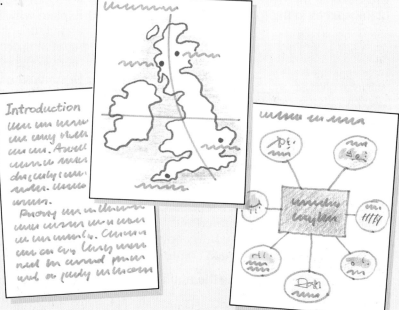

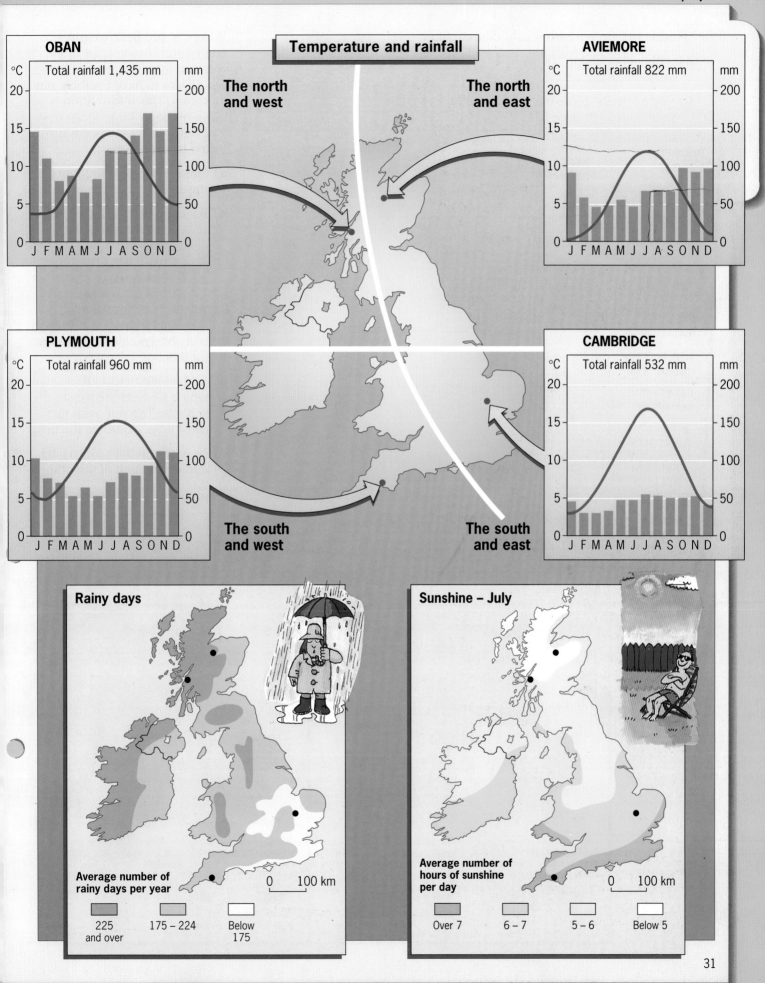

Temperature and rainfall

OBAN

Total rainfall 1,435 mm

°C ... mm

The north and west

The north and east

AVIEMORE

Total rainfall 822 mm

°C ... mm

PLYMOUTH

Total rainfall 960 mm

°C ... mm

The south and west

The south and east

CAMBRIDGE

Total rainfall 532 mm

°C ... mm

Rainy days

Average number of rainy days per year

| 225 and over | 175 – 224 | Below 175 |

0   100 km

Sunshine – July

Average number of hours of sunshine per day

| Over 7 | 6 – 7 | 5 – 6 | Below 5 |

0   100 km

2 What is Britain's weather and climate like?

a) Make a larger copy of the table of Britain's weather.

b) Complete the table using information from this page and from page 31.

Britain's weather	Oban (north and west)	Aviemore (north and east)	Plymouth (south and west)	Cambridge (south and east)
January temperature (°C)				
July temperature (°C)				
January rainfall (mm)				
July rainfall (mm)				
Total rainfall (mm per year)				
Rainy days (number per year)				
July sunshine (hours per day)				
Snow lying (days per year)				
Average wind strength (description and km/h)				

3 Where is the best weather?

Each family made a list of the weather and climate that they would like to have for their stay in Britain. This information is given on page 33. You can now find out which places are most suited to each family.

a) Make a copy of the table for the Jackson family.

b) For each place in turn put:
   ✓ a tick if the weather is suitable
   ✗ a cross if it is unsuitable
   ? a question mark if it is not perfect but not too bad.
   Your completed table showing Britain's weather will give you all the answers for this.

c) Add up the ticks to find which place is the most suitable. The one with the most ticks would be the best.

d) Now repeat the parts **a)**, **b)** and **c)** for each of the other three families.

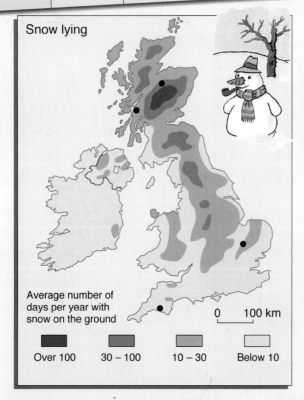

Snow lying

Average number of days per year with snow on the ground

0   100 km

Over 100   30 – 100   10 – 30   Below 10

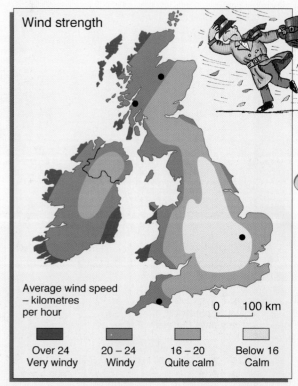

Wind strength

Average wind speed – kilometres per hour

0   100 km

Over 24 Very windy   20 – 24 Windy   16 – 20 Quite calm   Below 16 Calm

> We are a cycling family so we don't like rain or wind. We prefer warm summers and cold winters.

The Jackson family	Oban (north and west)	Aviemore (north and east)	Plymouth (south and west)	Cambridge (south and east)
Cold winters (Jan. temp. below 3°C)				
Warm summers (July temp. 15–20°C)				
Dry (less than 175 rainy days)				
Quite sunny in summer (6–7 hrs per day)				
Very little wind (below 16 km/h)				
TOTAL				

> We prefer it not to be cold or too snowy. We like to go fishing so rainy days can be good for us.

The Houston family	Oban (north and west)	Aviemore (north and east)	Plymouth (south and west)	Cambridge (south and east)
Mild winters (Jan. temp. 3–7°C)				
Mild summers (July temp. 10–14°C)				
Many rainy days (over 225 per year)				
A little snow (10–30 days per year)				
Windy (20–24 km/h)				
TOTAL				

> We are keen walkers and skiers. Our favourite days are in winter when it is cold and snowy.

> We enjoy barbecues and relaxing in the sun. We like warm sunny summers. Rain doesn't bother us but we really don't like the cold.

The Grant family	Oban (north and west)	Aviemore (north and east)	Plymouth (south and west)	Cambridge (south and east)
Cold winters (Jan. temp. below 3°C)				
Mild summers (July temp. 10–14°C)				
Quite dry (total rain 600–900 mm)				
Cloudy summers (under 5 hrs per day)				
Lots of snow (over 30 days per year)				
TOTAL				

The Stolberg family	Oban (north and west)	Aviemore (north and east)	Plymouth (south and west)	Cambridge (south and east)
Mild winters (Jan. temp. 3–7°C)				
Warm summers (July temp. 15–20°C)				
Quite wet (total rain 900–1,200 mm)				
Lots of summer sunshine (over 7 hrs per day)				
Windy (20–24 km/h)				
TOTAL				

4 Conclusion

Now you should look carefully at your work and answer the enquiry question. You could do this by replying to the letter from the World Wide Leisure Corporation. This could include writing and perhaps a labelled map.

a) Describe the weather and climate in each of the four regions of Britain. Your completed tables from this page will help you.

b) Say which place is best suited to each of the four families. Give reasons for your answer.

How does the water cycle work?

Rainy days are annoying and a nuisance to most of us. Yet rain is very important to our world as it is part of a never-ending cycle in which water is used over and over again. This cycle is called the **water cycle**. The amount of water in the cycle always stays the same. Some of the water may be **stored** in the sea, in the air or on land. Later, some of this water will be moved or **transferred** around the cycle. The main **stores** and **transfers** in the cycle are shown in diagram **A**.

The water cycle can be very complicated but its main features are shown in diagram **B**. Notice that the water can be moved in different forms – as vapour, rain, snow or hail. Some of the geographical terms used on this diagram are long and will be new to you. Chart **C** on the next page explains what these words mean.

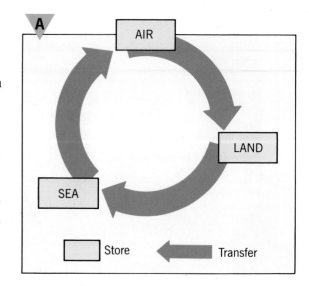

A

AIR

LAND

SEA

☐ Store   ← Transfer

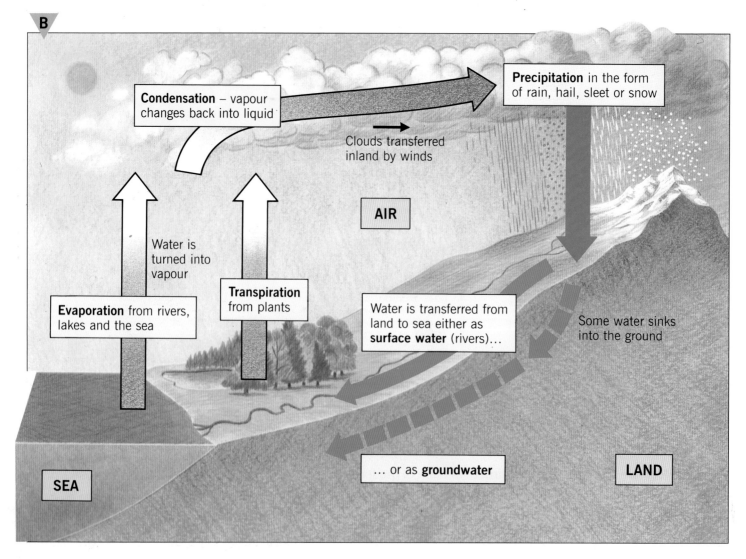

B

Condensation – vapour changes back into liquid

Clouds transferred inland by winds

Precipitation in the form of rain, hail, sleet or snow

AIR

Water is turned into vapour

Transpiration from plants

Evaporation from rivers, lakes and the sea

Water is transferred from land to sea either as **surface water** (rivers)...

Some water sinks into the ground

... or as **groundwater**

LAND

SEA

C

Evaporation	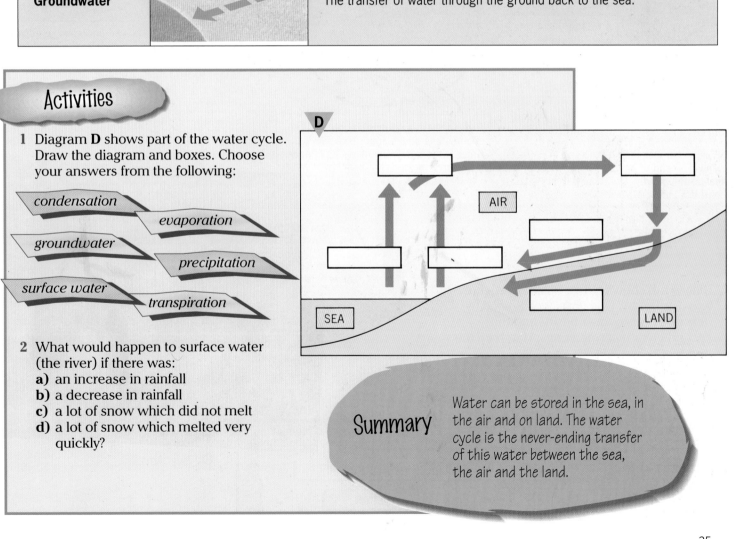	The transfer and change of water from the ground into water vapour in the air. Water vapour is an invisible gas.
Transpiration		The transfer and change of water from plants into water vapour in the air.
Condensation		Water vapour in the air changes back into a liquid. It forms small droplets which are visible as cloud.
Precipitation		The transfer of water from the air to the land. Water can fall to earth as rain, hail, sleet or snow.
Surface water		The transfer of water back to the sea over the ground surface. It is called surface run-off. It is easiest to see where it forms rivers.
Groundwater		The transfer of water through the ground back to the sea.

Activities

1 Diagram **D** shows part of the water cycle. Draw the diagram and boxes. Choose your answers from the following:

condensation

evaporation

groundwater

precipitation

surface water

transpiration

D

AIR

SEA

LAND

2 What would happen to surface water (the river) if there was:
a) an increase in rainfall
b) a decrease in rainfall
c) a lot of snow which did not melt
d) a lot of snow which melted very quickly?

Summary

Water can be stored in the sea, in the air and on land. The water cycle is the never-ending transfer of this water between the sea, the air and the land.

What is a river basin?

Most water that falls on land as rain eventually finds its way into a river. This is an important part of the **water cycle** which was explained on pages 34 and 35. Rain is a type of **transfer** whilst the river is a **store**. Although rivers differ in many ways, they all have similar features. Some of these are shown below.

If you look at the top of a large tree you will see lots of twigs. Twigs are small branches.

If you follow these downwards, you will see these twigs joining to form branches. These branches in turn join to form one big trunk.

A **river** is like a tree. It has lots of small streams which join to form **tributaries** which later join to form the main river. When it rains, most of the water slowly drains into streams, then into tributaries and finally into the main river. A **river basin** is the area of land drained by a main river and its tributaries.

A

Twigs

Bigger branches

Main trunk

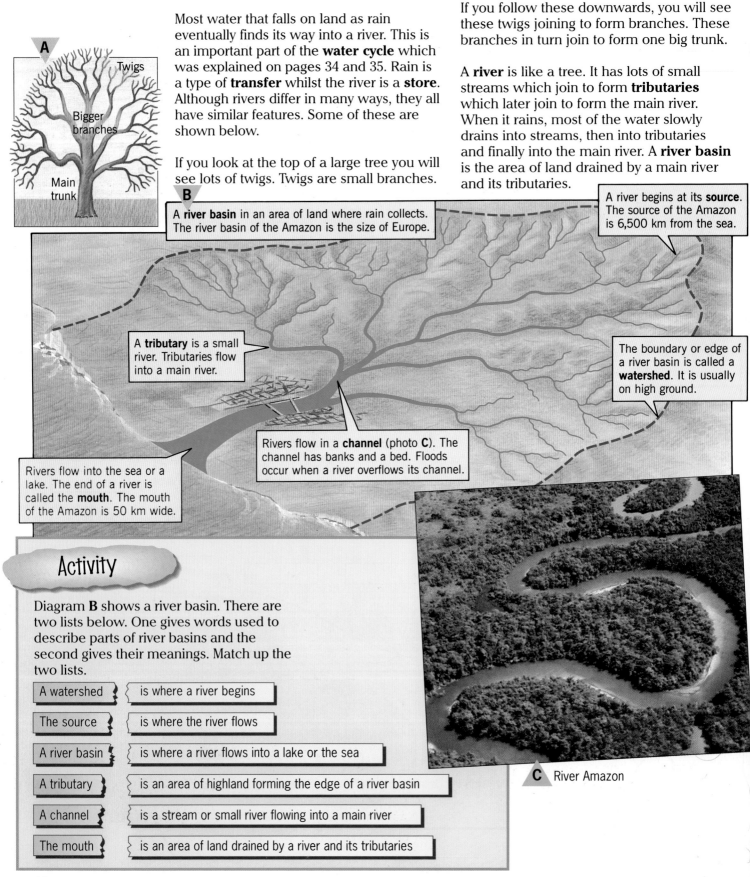

B

A **river basin** in an area of land where rain collects. The river basin of the Amazon is the size of Europe.

A river begins at its **source**. The source of the Amazon is 6,500 km from the sea.

A **tributary** is a small river. Tributaries flow into a main river.

The boundary or edge of a river basin is called a **watershed**. It is usually on high ground.

Rivers flow in a **channel** (photo **C**). The channel has banks and a bed. Floods occur when a river overflows its channel.

Rivers flow into the sea or a lake. The end of a river is called the **mouth**. The mouth of the Amazon is 50 km wide.

C River Amazon

Activity

Diagram **B** shows a river basin. There are two lists below. One gives words used to describe parts of river basins and the second gives their meanings. Match up the two lists.

A watershed	is where a river begins
The source	is where the river flows
A river basin	is where a river flows into a lake or the sea
A tributary	is an area of highland forming the edge of a river basin
A channel	is a stream or small river flowing into a main river
The mouth	is an area of land drained by a river and its tributaries

Where are the world's most important rivers?

Activities

Thirteen important world rivers are, in alphabetical order, the Amazon, Colorado, Danube, Ganges, Murray-Darling, Mississippi, Nile, Rhine, St Lawrence, Volga, Yangtze, Zambezi and Zaire (Congo).

1 Fit the names of these rivers into crossword **D** on the right. The number of letters in each word will help you to fit them into the puzzle. For example, the Murray-Darling, which has been done for you, was the only word with 13 letters. It could only fit into that one place.

2 With help from map **E** below, sort the rivers into groups under the headings: *Africa, America, Asia, Australia* and *Europe*.

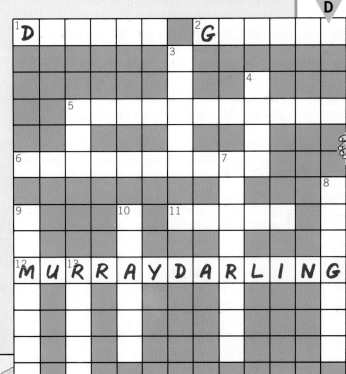

D

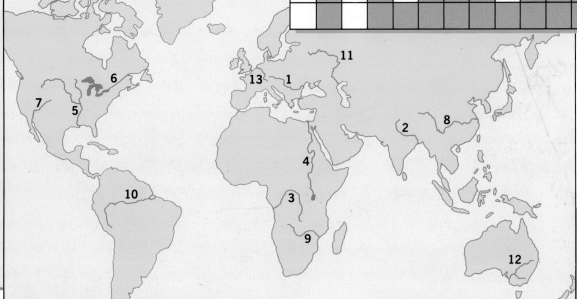

E

Summary

Rain collects in rivers in a river basin. Rivers have their source in highland areas and flow in a channel to the sea or to a lake.

E X T R A S

1 The three longest rivers in the United Kingdom are the Severn, the Thames and the Trent. Name **one** city on each of these rivers.

2 Match the 13 rivers named in question **1** above with a country through which they flow. Some flow through several countries, so choose only **one** important country.

What causes a river to flood?

All of the water that flows down a river comes from rain or melting snow. Sometimes after heavy rain or a rapid snow melt, there may be too much water for the river to hold. The river will then overflow its banks and spread out across the land on either side of its channel. This is called a **river flood**.

Usually when it rains, most water simply soaks into the ground and there is little chance of a flood. If, however, the water is unable to soak into the ground, it will stay on the surface and flow quickly downhill and into the river. This is when floods are most common.

Some rivers are more at risk from flooding than others. Put simply, heavy rain and anything which stops that rain from soaking into the ground will increase the chances of flooding. Some of the factors that increase the risk of flooding are shown below.

A The rapid melting of snow can cause flooding

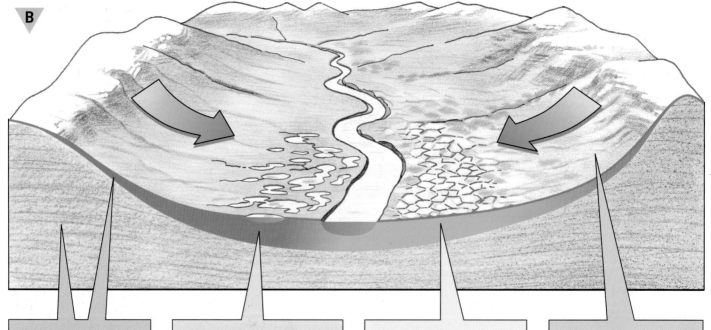

B

Rock and soil type
Impermeable rocks and soils do not allow rain to soak through them. Any rain that falls will stay near the surface.

Very wet soil
If rain has been falling for some time, the soil may become full of water. Any further rain is unable to soak into the ground and remains on the surface.

Very dry soil
Soil that is baked hard by the sun in dry weather builds up a crust. Rain is unable to soak through the crust and so remains on the surface.

Steep slopes
Rain falling on a steep slope runs quickly downhill towards a river. It has little time to soak into the ground, so most stays on the surface.

Floods are more common now than they used to be. There are more of them and they are increasing in size. Many people are blaming human activity for this.

Two ways in which humans may increase the risk of flooding are by cutting down trees and building more towns and cities. These are shown in drawings **C** and **D**.

C

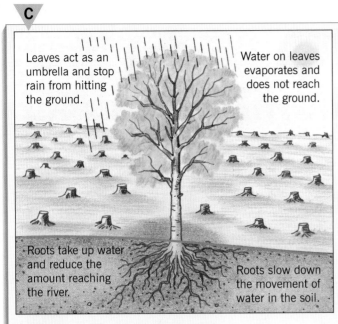

Leaves act as an umbrella and stop rain from hitting the ground.

Water on leaves evaporates and does not reach the ground.

Roots take up water and reduce the amount reaching the river.

Roots slow down the movement of water in the soil.

Cutting down trees (deforestation)
Many of the world's forests are being cleared to make way for other developments. In some countries the number of serious floods has more than doubled since large-scale tree clearing began.

D

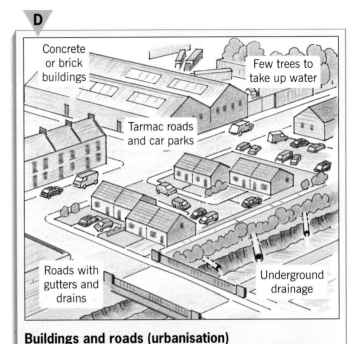

Concrete or brick buildings

Few trees to take up water

Tarmac roads and car parks

Roads with gutters and drains

Underground drainage

Buildings and roads (urbanisation)
Rain falling on concrete and tarmac is unable to soak into the ground, so stays on the surface. Gutters and drains then carry the water quickly and directly to the river. Large towns are most at risk.

Activities

1 a) Make a larger copy of drawing **E**.
   b) Add the following labels to your drawing to show how a river floods:
   ● River level rises
   ● Water quickly reaches river
   ● River floods
   ● Water runs over surface
   ● Heavy rain falls
   ● Rain soaks into ground

2 Describe four factors that increase the risk of flooding.

3 With the help of diagrams, describe how:
   a) cutting down trees, and
   b) building large towns can make floods worse.

F

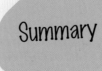

E

Summary River flooding is most likely after heavy rain or rapid snow melt. The flood risk is greatest when water is unable to soak into the ground. Human activities can increase the chance of flooding.

Floods in the UK, 2000

Evening **COURIER**

Sussex
Friday, 13 October 2000

FLOOD HAVOC HITS SOUTHERN ENGLAND

A massive rescue operation was underway last night as tens of thousands of homes were hit by the UK's worst floods for 30 years.

More than a month's rain fell in 24 hours as torrential downpours and storms caused six of this region's rivers to burst their banks. Thousands of homes, shops and offices filled with water up to 2 metres deep. Most of the main roads in the region were blocked and all rail services cancelled.

Emergency services, including lifeboats, coastguard helicopters, the fire service and police were scrambled to rescue hundreds of people trapped in buildings and on rooftops.

Worst hit was Uckfield which was completely submerged and cut off after a wall of water, almost 2 metres in height, crashed through the town centre. Cars were swept away and the soggy contents of shops floated into the streets. Thousands of homes in the area suffered severe damage. Many people have lost all of their belongings. The council have set up refuge centres at a local school and leisure centre.

As the flood water subsides, people returning home are finding everything covered in a thick layer of foul-smelling mud. The clean-up operation will take weeks. Many people will not be back in their homes for months and some businesses may never re-open.

Insurance experts are putting the cost at over £500 million but there are fears that many people will not be insured. The government has promised help to areas most in need.

The south of England has had many floods recently. Most have been caused by very heavy rain. Floods like these are called **flash floods** because they happen very suddenly and last for just a short time. Flash floods can be dangerous as they come without warning and give people little time to escape.

A serious flood rarely has one single cause, though. In the south of England much of the land is low-lying and many new developments have been built on the floodplain. Some river embankments collapsed, and reservoirs that had reached the point of overflowing added to the problem.

A

The October 2000 flood

1 Heavy rain had been falling in the area for more than a week.

2 The ground became full of water and could take no more.

3 Almost two months of rain fell on the area in less than 24 hours.

4 Rivers burst their banks and water flooded surrounding areas.

5 Recent building in floodplain areas made the problem worse.

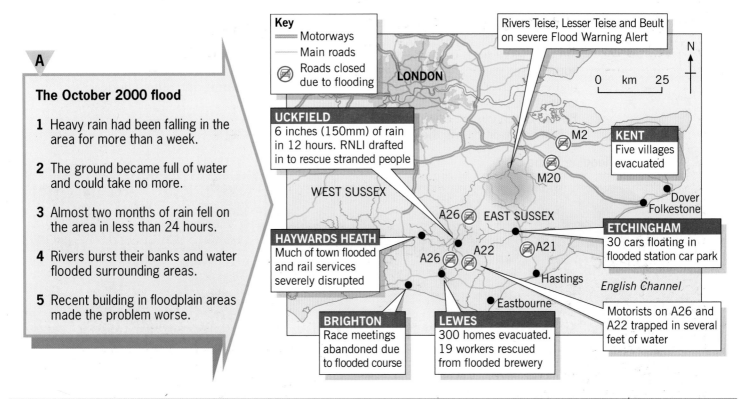

Key
- Motorways
- Main roads
- Roads closed due to flooding

Rivers Teise, Lesser Teise and Beult on severe Flood Warning Alert

LONDON

0   km   25   N

UCKFIELD
6 inches (150mm) of rain in 12 hours. RNLI drafted in to rescue stranded people

WEST SUSSEX

KENT
Five villages evacuated

Dover
Folkestone

HAYWARDS HEATH
Much of town flooded and rail services severely disrupted

EAST SUSSEX

ETCHINGHAM
30 cars floating in flooded station car park

A26   A22   A21

Hastings

English Channel

Eastbourne

BRIGHTON
Race meetings abandoned due to flooded course

LEWES
300 homes evacuated. 19 workers rescued from flooded brewery

Motorists on A26 and A22 trapped in several feet of water

M2   M20

Activities

1 Imagine that you are a reporter for the *Evening Courier* and have been asked to write a report on the flood disaster for the weekend paper. Make notes for your report using the headings below.

B

Flood disaster report

a) When did it happen?
b) Which places were worst affected?
c) What were the main effects of the flood?
d) How were car owners and rail travellers affected?
e) What help was available to flood victims?
f) What problems will there be once the flood waters have gone down?

2 Make a larger copy of diagram **C** and add six causes of the October 2000 flood.

C

Southern England: causes of flooding

Summary

Floods can cause much damage and seriously affect people's lives. There are usually several different causes of floods but some places are more at risk from flooding than others.

41

How does the UK cope with floods?

Flooding is a serious problem in the UK and it is happening more often. There are 1.3 million properties at risk and this number is expected to increase to over 5 million in the next 50 years. Autumn 2000 was the wettest since records began in 1766. Major flooding affected large parts of the country, and in some cases water levels were at their highest for over 100 years. Whilst flooding cannot be prevented, in rich countries like Britain much can be done to reduce the risk of floods and limit their worst effects.

The **Environment Agency** is an organisation that looks after rivers in England and Wales. It monitors rainfall, river levels and sea conditions 24 hours a day. This information is used to predict the possibility of flooding. If flooding is forecast, the Environment Agency's **Floodline** issues warnings. It also gives advice on what to do before, during and after a flood.

A Planning for flooding in the UK

① Study the UK's rivers and coasts and identify areas most at risk and where flooding would do most damage.

② Recommend the building of flood defences such as embankments and overflow channels where they are needed.

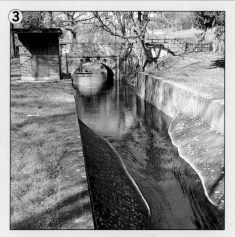

③ Continually check rainfall and water levels to see if a river is going to flood.

④ Issue warnings through radio, TV and home visits for those in most danger, when floods are expected.

⑤ Alert emergency services such as the police, fire brigade and army, to provide help for those in need.

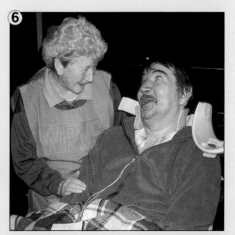

⑥ Ensure that food and shelter is available for those made homeless. Emergency medical care should also be available.

"Flooding – you can't prevent it, but you can prepare for it."

ENVIRONMENT AGENCY

Flood warning codes

 Flood Watch — **Flood Watch** means flooding is possible. Be aware! Be prepared! Watch out!

 **Flood Warning** — **Flood Warning** means flooding of homes, businesses and main roads is expected. Act now!

Severe Flood Warning — **Severe Flood Warning** means serious flooding is expected. There is imminent danger to life and property. Act now!

All Clear — **All Clear** means there are no longer flood watches or flood warnings in force. Seek advice to return.

What to do in a flood

Before a flood
Be alert for flood warnings and take action.
Check on family and nearby neighbours.
Move people, pets and valuables to safety.
Collect warm clothes, food and a torch.
Block doorways with sandbags.
Switch-off electricity and gas.

During a flood
Listen to the local radio for flood news.
Never walk, drive or swim through flood water.
Avoid flood water as it may be contaminated.

After a flood
Check if it is safe to turn electricity and gas on.
Open windows and doors for ventilation.
Throw out contaminated food.
Wash taps and run them before use.
Clear up by disinfecting walls and floors.
Beware of rogue traders offering to help.
Call your insurance company for advice.

Activities

1 **a)** Describe three ways that the Environment Agency can help reduce the risk of flooding.
   b) Describe three ways that the Agency can help limit the worst effects of flooding.

2 Which flood warning would have been given for the south of England floods of June 2000 (pages 40 and 41)? Give reasons for your answer.

3 'Floodline' encourages people to make a family flood plan like the one below. Write out the plan and add a reason for each point.

4 **a)** Find out more about Floodline by telephoning for a Flood Warning Pack or using their website.
   b) Design a leaflet to give to people living in areas where there is a flood risk. Explain to them briefly what information is available and what they should do.

ENVIRONMENT AGENCY
Floodline
Telephone: 0845 988 1188
Website: www.environment-agency.gov.uk/flood

Family Flood Plan
* Know how to contact each other.
* Put together an emergency flood kit.
* Know how to turn off power supplies.
* Put emergency numbers in a safe place.
* Understand the flood warning system.
* Listen to the local radio programme.

Summary

There is no easy way to cope with floods. Rich countries like the UK can afford schemes that help reduce the damaging effects of flooding.

Floods in Bangladesh, 1998

SURMA NEWSWEEKLY

**Dhaka
15 September 1998**

Worst ever floods hit Bangladesh

The heaviest rains in living memory have left a trail of destruction across Bangladesh. Estimates suggest that over 7 million homes have been destroyed and at least 25 million people made homeless. The official death toll is 2,379 although many more are still missing.

More than 80% of the country is covered in water. In some places only the tops of trees and buildings can be seen. Railways and roads have been swept away and Dhaka airport is still under water. Delivering emergency food and medical care to those in need has been almost impossible.

Few places have any electricity and there is no safe drinking water. The threat of disease is increasing and hospitals are already full of people suffering from dysentery and diarrhoea. Many of these people will not survive.

The countryside areas have been worst hit. One family, sheltering on the corrugated iron roof of their flooded home, had lost their two oldest children to the flood. Their mother was in tears.

"They were just washed away in the night and never seen again. We have lost our homes, lost our land and lost our cattle. Our crops have been ruined and we have no food or money. Without help we will starve. Please help us..."

What caused the Bangladesh flood?

Bangladesh is a country in Asia. It is located at the mouth of two of the world's longest rivers, the Ganges and the Brahmaputra.

Bangladesh has floods every year – but they seem to be getting worse. The country relies on the heavy monsoon rains to flood the rice fields, but too much rain can destroy the crop as well as the homes of the farmers. In four monsoon months, Bangladesh can get as much rain as London gets in two years!

A Some causes of flooding in Bangladesh

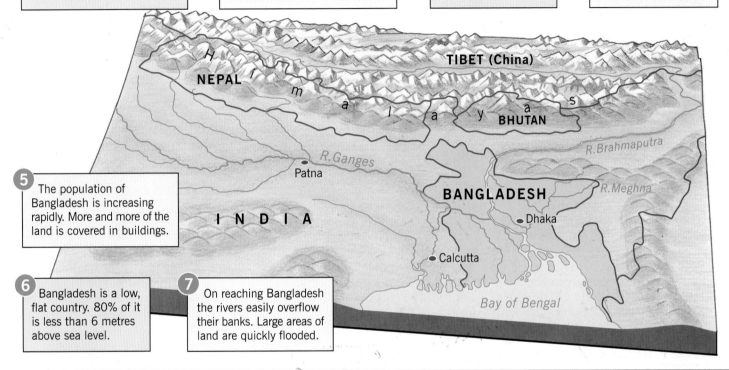

1 The Himalayas and Bangladesh get heavy monsoon rain. The last few years have been wetter than usual.

2 The number of people living in India and Nepal is increasing. Trees have been cleared to make way for farmland and housing.

3 The loss of trees has increased the amount of rainwater reaching the rivers.

4 The huge rivers bring the water quickly towards Bangladesh.

5 The population of Bangladesh is increasing rapidly. More and more of the land is covered in buildings.

6 Bangladesh is a low, flat country. 80% of it is less than 6 metres above sea level.

7 On reaching Bangladesh the rivers easily overflow their banks. Large areas of land are quickly flooded.

Activities

1 Describe the effects of flooding in Bangladesh using the newspaper headlines shown in **B**.

2 Flooding is due mainly to natural events. Explain how each of the following facts is a cause of floods in Bangladesh:
   a) monsoon rains fall from May to September
   b) many rivers flow into Bangladesh
   c) Bangladesh is a very flat country.

3 Explain how each of the following has made the flooding problem worse in Bangladesh:
   a) human activity in India and Nepal
   b) human activity in Bangladesh.

B

Death toll soars as millions made homeless

Transport links broken in worst flood of all

Starvation and disease as floods drown country

Families worst hit as floods wreck homes

Summary Flooding can seriously affect people's lives. Floods may result from natural events or from human activity.

How does Bangladesh cope with floods?

Bangladesh suffers more from flooding than any other country in the world. The problem is made worse because of the extreme poverty of the people who live there.

In 1989, after a particularly bad flood, several wealthy countries joined with Bangladesh to set up the Flood Action Plan. Under the Plan, billions of dollars are being spent on schemes which it is hoped will reduce the risk and danger of flooding. Some of the main points of the Plan are shown below.

A **Flood Action Plan for Bangladesh**

Build **dams** to control river flow and hold back the monsoon rainwater in reservoirs. Stored water can be used for irrigation and to generate cheap electricity.

Build **embankments** and deepen river channels to stop the river overflowing. The embankments would be up to 7 metres high in urban areas.

Build 5,000 **flood shelters** in areas most at risk. These would be cheap to construct and provide a place of safety for almost everyone. They would be well stocked with food.

Improve **flood warning systems**. These would give early warnings of floods. They would also give instructions to people as to what they should do before, during and after the flood.

Provide **emergency help** when the floods arrive. Embankments would be repaired, people taken to safety and food and medical care provided to those in need.

Give **after-care** once the flood ends. Food, drinking water, tents, medicines and money would be available. Help would be given to plant seeds for next year's crops.

There is no easy solution to Bangladesh's flooding problem. The enormous size of the problem and the shortage of money make the task almost impossible. Even the Flood Action Plan has not been welcomed by everyone. Many people are worried that such a large scheme could actually make the problem worse.

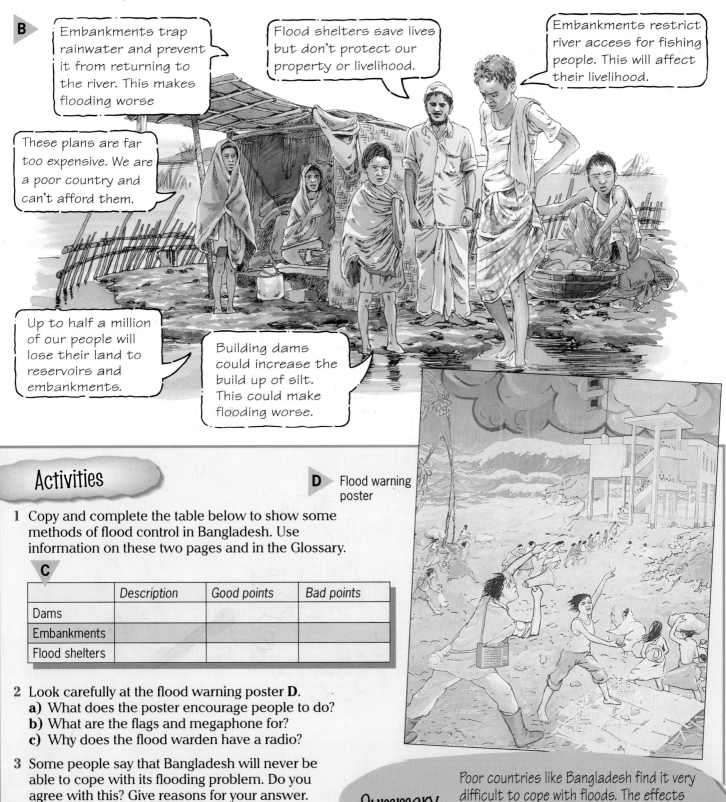

B

> Embankments trap rainwater and prevent it from returning to the river. This makes flooding worse

> Flood shelters save lives but don't protect our property or livelihood.

> Embankments restrict river access for fishing people. This will affect their livelihood.

> These plans are far too expensive. We are a poor country and can't afford them.

> Up to half a million of our people will lose their land to reservoirs and embankments.

> Building dams could increase the build up of silt. This could make flooding worse.

D Flood warning poster

Activities

1 Copy and complete the table below to show some methods of flood control in Bangladesh. Use information on these two pages and in the Glossary.

C

	Description	Good points	Bad points
Dams			
Embankments			
Flood shelters			

2 Look carefully at the flood warning poster **D**.
   a) What does the poster encourage people to do?
   b) What are the flags and megaphone for?
   c) Why does the flood warden have a radio?

3 Some people say that Bangladesh will never be able to cope with its flooding problem. Do you agree with this? Give reasons for your answer.

Summary

Poor countries like Bangladesh find it very difficult to cope with floods. The effects of flooding are therefore a lot worse than they would be for a rich country.

How can the risk of flooding be reduced?

There are many different ways of controlling rivers and reducing the risk of flooding. The methods shown below are called **flood prevention schemes** because they try to stop floods happening.

Many people now believe that complete river and flood control is impossible. They say that flooding should be allowed to happen as a natural event. Flood prevention schemes can, in the long term, save money. They also improve water quality and help support wildlife.

A

Dams
A dam built across a river traps water and stores it in a reservoir. The water may then be released in a controlled way.

Forests
Trees may be planted in the drainage basin. These will slow down water movement and reduce the amount reaching the river.

Embankments
The river's banks may be built up with earth or concrete. This will make the river deeper and keep the water in.

Concrete linings
Line river channels in urban areas with concrete. This will take excess water quickly away from danger areas.

Activities

1 Draw a star diagram to show eight ways of reducing the risk of flooding. Write a short sentence to describe each one.

2 Look at the different approaches to flood prevention. Which approach do you think:
   a) costs most
   b) costs least
   c) may drown farmland and houses
   d) uses up most land
   e) protects the natural environment?
   Give reasons for your answers.

3 One approach to flooding is simply to allow rivers to flood naturally. For each of the people below say if they would be **for** or **against** this method. Give reasons for your answer.

Local farmer   Flood protection manager   Bird watcher

Summary

A variety of methods can be used to reduce the risk of floods, but there is no way to stop flooding completely. A modern approach is to allow parts of a river to flood naturally.

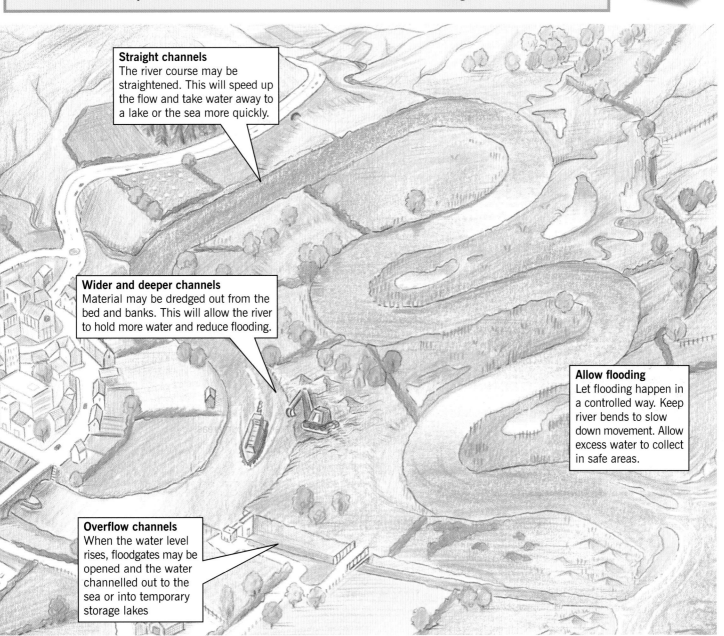

Straight channels
The river course may be straightened. This will speed up the flow and take water away to a lake or the sea more quickly.

Wider and deeper channels
Material may be dredged out from the bed and banks. This will allow the river to hold more water and reduce flooding.

Allow flooding
Let flooding happen in a controlled way. Keep river bends to slow down movement. Allow excess water to collect in safe areas.

Overflow channels
When the water level rises, floodgates may be opened and the water channelled out to the sea or into temporary storage lakes

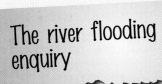

The river flooding enquiry

How should the Doveton valley be protected from flooding?

Look at map **D**, which shows part of the Doveton valley. In most years, the river overflows its banks and causes serious damage. The people in the area want their homes and land to be protected from the flooding. The Environment Agency has agreed to look at the problem. It has made a study of the area and suggested four different schemes to help stop the flooding. It is your task to decide which is the best scheme.

A

Factors to consider	Scheme A	Scheme B	Scheme C	Scheme D
Prevents all flooding				
Stops flooding in Crofton				
No homes lost				
No roads submerged				
No grazing land lost				
No good farmland lost				
Helps with irrigation				
Helps protect wildlife				
Not too expensive				
Total				

B

1 a) Copy table **A** which shows some factors that have to be considered when choosing a flood protection scheme.

b) Look carefully at the map and scheme descriptions. Show the advantages of each scheme by putting ticks in columns A, B, C or D. Complete one factor at a time. More than one column may be ticked for each factor.

c) Add up the ticks to find which scheme has the most advantages.

d) Which scheme would *you* choose? The one with the most advantages would be the best. If two schemes are equal, think about which parts of the valley you would want to protect most.

e) Briefly describe the scheme you have chosen. Explain how it will help protect the valley from flooding.

2 The flood protection scheme will affect different people in different ways. Work in pairs and discuss what the people in the drawing below will think of your chosen scheme. For each person say if they would be **for** or **against** the scheme. Give reasons for their views.

C

Trudy Trout, owner of Crofton caravan Park

Farmer Wally Wade of Hillside Farm

Barry Beer, owner of the Crofton Inn

Larry Laugh, local lorry driver

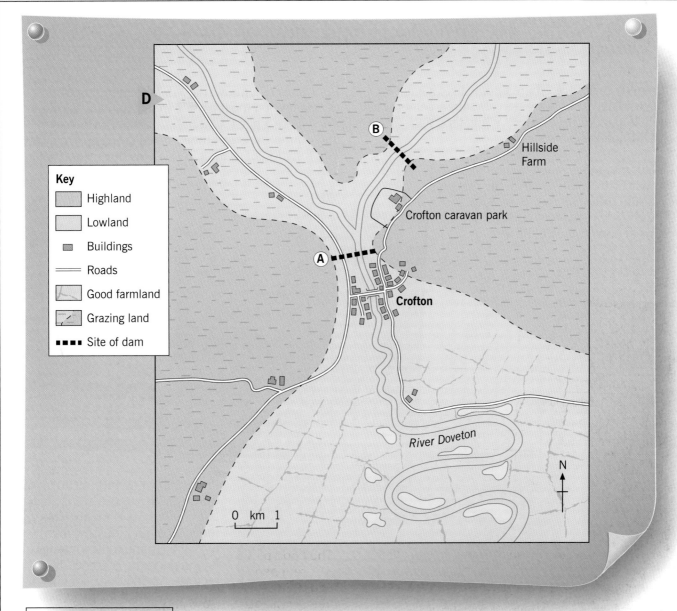

Very expensive = **££££**
Quite cheap = **£**

Scheme A

Build a dam at A and create a large reservoir above the village. Much farmland and several farms would be flooded. The scheme would stop flooding in the village and protect most of the valley. **Cost = ££££**

Scheme B

Build a dam at B and create a small reservoir higher up the valley. Nobody would lose their home but some grazing land used by sheep and cattle would be lost. There would still be some flooding in Crofton and further downstream. **Cost = ££**

Scheme C

Build a dam at B and deepen the river channel through Crofton. This would allow the water flowing through the village to move away more quickly. The scheme would protect Crofton but there may still be some flooding downstream. **Cost = £££**

Scheme D

Build embankments at Crofton. Deepen and straighten the river below the village to take water away quickly. Allow natural flooding to happen downstream at the river bends. There would still be some flooding, especially upstream of Crofton. **Cost = £**

What are settlements and why do they grow?

A **settlement** is a **place** where people live. It can be as small as a village or as large as a city. Many settlements in Britain began, or **originated**, a long time ago – some even in Roman times.

For a settlement to grow there had to be a special reason why it should be built in the first place. This is called a **function**. The four photos on these two places show four different functions of settlements in Britain.

Photo **A** is of a **market** town. Market towns were needed when most people in Britain were farmers. They were a place where farmers could buy seed, tools and animals and sell their grown crops and animals. The function of market towns was buying and selling.

> The market is in a square. Around the square are banks, pubs, shops and offices used by farmers and other people when they visit the market.

Photo **B** is of an **industrial** town. Industrial towns grew up much later than market towns. The function of an industrial town was to make (manufacture) things in factories. Some of the early factories used goods sold by farmers, such as wool, to make things. Other factories made things needed by farmers, such as machines.

> The tall chimneys and large buildings belong to factories. Factories are found in industrial (manufacturing) towns.

The towns in photos **C** and **D** both grew up to help and serve people living in market and industrial towns.

Photo **C** shows a **port**. It brings in goods from overseas countries which are needed on local farms and in factories. Later the port will send products from these farms and factories back overseas. Goods or products that are brought into a country are called **imports**. Goods that are sent overseas are called **exports**.

Photo **D** shows a **holiday resort**. People from nearby settlements come here to relax and enjoy themselves.

Some settlements still have their original function but others have changed. Dover had a castle to try to stop people invading Britain. Now it is a port to help people come into Britain. Larger settlements, like London, have had more than one function, e.g. market, port, industry and now as a centre for holidays and business.

> A harbour is a sheltered place for ships. A port is where the harbour is surrounded by docks, warehouses, roads and railways.

> People go to holiday resorts like this one to swim in the sea and relax on the beach. They stay in hotels and need things to do (entertainment).

Activities

1 Complete the following definitions by matching the beginnings on the left with the correct ending from the list on the right.

A settlement	is where people buy and sell things
A function	is where goods are brought into and sent out of a country
A market town	is a place where people live
An industrial town	is where people go on holiday to relax
A port	is where people make things
A resort	is the reason why a town was first built

2 **a)** What was the original function of your local town or city?
  b) What are its functions today?

Summary

For a settlement to grow up (originate) in the first place, it had to have a special use (function). Different settlements grew up with different functions.

How were the sites for early settlements chosen?

When we use the word **site** we mean the actual place where a village or town grew up. A site was chosen if it had one or more natural advantages. Diagram **A** shows eight natural advantages. The more natural advantages a place had the more likely it was to grow in size.

A

Protection. Good views from a hilltop give you warning if you are about to be attacked.

Plenty of water. Needed for drinking, cooking and washing. Water might come from a river, spring or well.

Not too much water. Sites must not flood or be marshy.

Rivers. Easy to cross either on foot at a ford or by a bridge.

Building materials. Needed wood or stone. Useful to be near a wood or a rocky hillside.

Supply of wood. Needed for fires for warmth and to cook on.

Shelter. A south facing slope will have more sun and will be protected from the cold north wind.

Flat land. Easier to build on, for growing crops and travelling to other towns.

'Is this a good place to build a village?'

'Is this a good place to build a town?'

Activities

1 Write down the meaning of the word 'site'.

2 Landsketch **B** shows an area in Ancient Britain. On it, labelled **A**, **B**, **C**, **D** and **E**, are five possible sites for a village.
   a) Suggest at least **one** natural advantage of each site.
   b) Suggest at least **one** natural disadvantage of each site.
   c) Which site would you choose? Give **three** reasons for your choice.

3 Try to find out what were the natural advantages of the site of your own town or village.

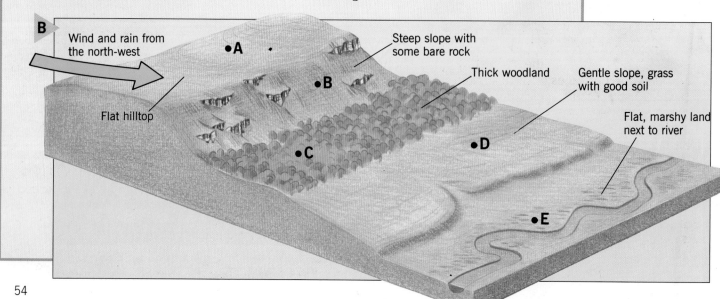

B

Wind and rain from the north-west

Steep slope with some bare rock

Thick woodland

Gentle slope, grass with good soil

Flat hilltop

Flat, marshy land next to river

•A
•B
•C
•D
•E

The photo below shows Warkworth, a small village in Northumberland. It is located on a bend of the River Coquet. Early settlers were most concerned about their safety and getting a supply of food and water. As you can see, the site at Warkworth provided those needs.

Despite its good site, Warkworth has never grown into a large town. This is because the original advantages are no longer so important. Nowadays, people want to be near employment and services such as schools and hospitals. These are not so readily available at Warkworth.

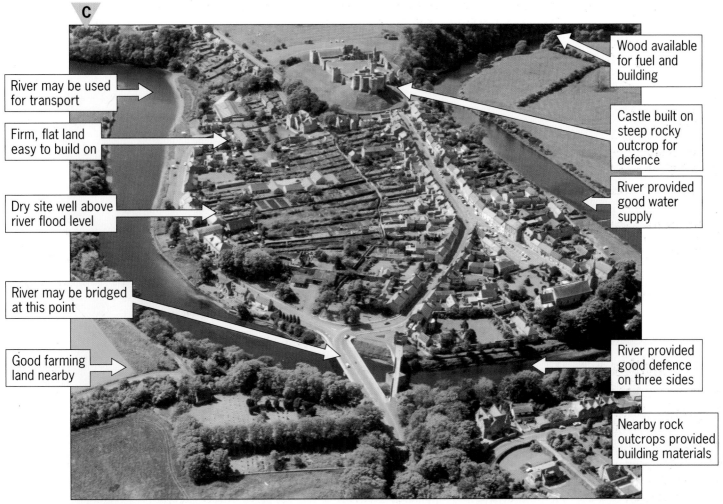

C

River may be used for transport

Firm, flat land easy to build on

Dry site well above river flood level

River may be bridged at this point

Good farming land nearby

Wood available for fuel and building

Castle built on steep rocky outcrop for defence

River provided good water supply

River provided good defence on three sides

Nearby rock outcrops provided building materials

Activities

1 Draw a star diagram like the one below to show the advantages of Warkworth as a site for a settlement. Give two advantages under each heading.

Defence

Food and water

Site advantages of Warkworth

Building materials

Building land

2 Complete table **D** to show how **some** of Warkworth's original site advantages are no longer so important.

D

Original advantage	Why no longer important
•	
•	

Summary

Early sites for settlements were chosen because of natural advantages such as a good water supply, dry land, defence, shelter, farmland and building materials.

What different settlement patterns are there?

If you look at map **D** on the next page you will see that the settlements have different shapes. Some are long and thin, some are compact and almost round, others are broken up and spread out. In geography we call these different shapes the **settlement pattern**.

Settlement patterns are usually influenced by the natural features of the area. These are often the same features that were considered important when choosing the original site for the settlement. The three main types of settlement pattern are shown below.

- A **dispersed settlement** has buildings that are well spread out.

- Settlements with this pattern are often found in highland areas where it is not easy to build houses close together. Here, people also needed more land to grow their crops or graze their animals.

Dispersed

A

- A **nucleated settlement** has buildings closely grouped together.

- Settlements with this shape often grew around a road junction or river crossing. A long time ago people built their houses close together for safety. This pattern is common in lower, flatter parts of Britain.

Nucleated

B

- **Linear settlements** are often called **ribbon developments** because they have a long, narrow shape.

- Settlements with this shape usually grow along a narrow valley where there is little space. They may also be found strung along a road or on either side of a river.

Linear

C

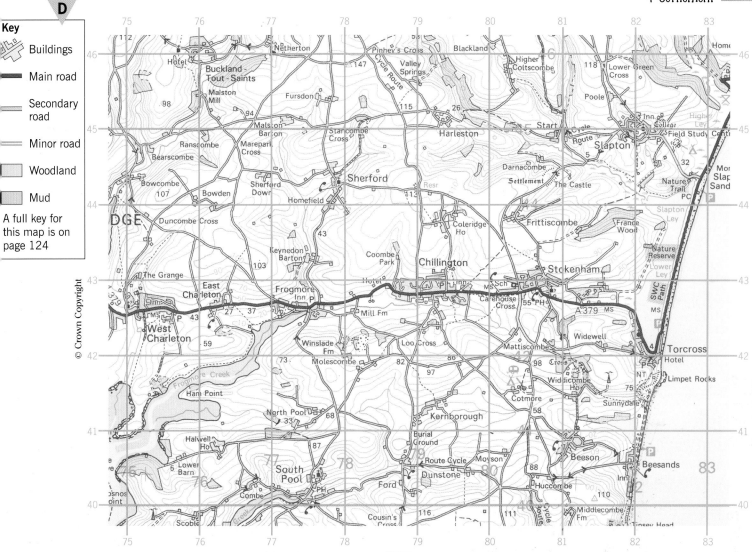

Key

- Buildings
- Main road
- Secondary road
- Minor road
- Woodland
- Mud

A full key for this map is on page 124

© Crown Copyright

Activities

1 Copy the settlement pattern drawings below. Label each one **dispersed**, **nucleated** or **linear**. Write a brief description of each one. Suggest a reason for its shape.

 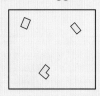 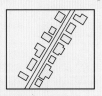

E

2 Map **D** is part of Devon in south-west England. It shows many different settlement patterns.

a) Make a larger copy of table **F**.

b) Complete your table by filling in the empty boxes. The first one has been done to help you. (You may need to look at page 114 to remind you about grid references.)

c) Find another example of a nucleated settlement and a linear settlement and add them to your table.

F

Village name	Map reference	Simple drawing	Settlement pattern
Bowden	7644		Dispersed
Slapton	8144		
North Pool	7741		
Cotmore	8041		
Beeson			
Torcross			
Sherford			

Summary

The three main types of settlement pattern are dispersed, nucleated and linear. The shape of a settlement is usually determined by the physical features of the surrounding area.

What are the benefits and problems of settlement growth?

In Britain most people are urban dwellers living in towns and cities. These settlements grew very quickly in the nineteenth century. This was when large numbers of people moved there to work. Today, Britain's cities are no longer growing in size. However, in many overseas countries people are still moving to cities in large numbers.

This is because they believe that many **benefits** come from living and working in cities. Moving there will improve their **quality of life**.

Drawing **A** shows some of the benefits which people hope to find in large towns and cities.

A

- There are more houses and flats to buy or to rent.
- There are more jobs which are better paid.
- Food supplies are more reliable, with many shops giving a greater choice.
- It takes less time and money to travel to work and to shops.

- There are more and better services, such as schools and hospitals.
- Urban areas have 'bright light' attractions, such as discos, concerts and sporting activities.

For people already living in cities, life is often less attractive. Living and working in cities creates many **problems**. Drawing **B** shows some of the problems found in British cities.

B

Old housing

Disused factories

SITE TO LET

Pollution

Traffic congestion, pollution and noise

High prices

Crime and vandalism

JOB CENTRE VACANCIES

Litter and rubbish dumping

Unemployment

Homelessness

- Traffic causes congestion, accidents, noise and air pollution.
- Old roads are too narrow for lorries and buses; new roads take up much land.
- Old houses and factories need urgent, expensive repairs or they are left empty.

- There is waste land where houses and factories have been pulled down.
- Crime, vandalism and litter make cities dangerous and unpleasant.
- Land is very expensive to buy, in and near the city centre.

Activities

1 a) Make a copy of the table below. List the **three** things which you think are best about living in cities, and the **three** things you think are the worst.

Cities	
Good news	Bad news

b) Do you think there is more good news or bad news?

2 If you had to move from where you live, would it be to a bigger or a smaller settlement? Give reasons for your answer.

E X T R A S

1 Try to find out what has been done in your local town or city to try to reduce
   a) traffic problems
   b) pollution
   c) crime, vandalism and litter.

2 Suggest other ways in which these three problems may be overcome.

Summary

Many people move to large cities because they see benefits in living and working there. However, as these settlements become older and bigger, many problems are created.

How do settlements change with time?

No town or village remains the same for ever. Over a period of time the following may all change:

1 the **shape** of a settlement
2 the **function** of a settlement
3 the **land use** of a settlement
4 the **number** and **type** of people living in the settlement.

Villages are small in size so it is often easier to see these changes in them than it is to see changes in a large town or city.

What was a typical village like in the 1890s? Although no two villages are the same, most have several things in common. Diagram **A** shows a typical village about one hundred years ago. In the centre there was often a village green. Buildings were grouped closely together (nucleated) around this green forming a **core**. Roads were usually narrow lanes. Most houses were small terraced cottages. The people who lived in them would probably have been born in the village. Most would have worked on local farms. As houses and farms were built at different times they would have different styles and building materials.

How has the village changed by the 1990s? Diagram **B** shows the same village today. The village has grown larger and has many new buildings. It has become **suburbanised**. This means it has become similar to the outskirts of larger towns.

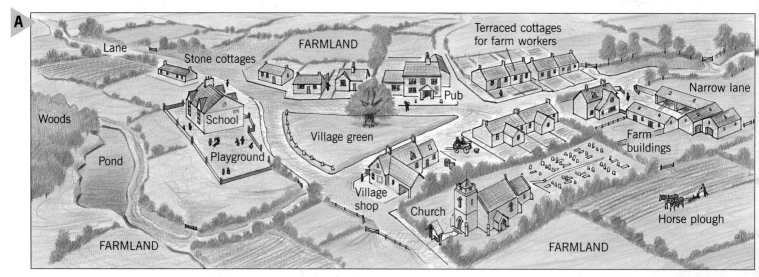

A — Lane; Stone cottages; FARMLAND; Terraced cottages for farm workers; Narrow lane; Woods; School; Village green; Pub; Farm buildings; Pond; Playground; Village shop; Church; Horse plough; FARMLAND; FARMLAND

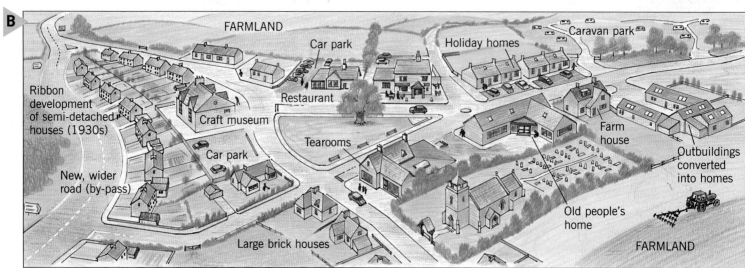

B — FARMLAND; Car park; Holiday homes; Caravan park; Ribbon development of semi-detached houses (1930s); Restaurant; Craft museum; Tearooms; Car park; Farm house; Outbuildings converted into homes; New, wider road (by-pass); Old people's home; Large brick houses; FARMLAND

Activities

1 Write down the meaning of
   a) shape
   b) function
   c) land use
   when talking about a settlement.

2 **Spot the differences!** List at least **ten** differences between the village in the 1890s and the village in the 1990s.

3 The changes to the village will have affected different groups of people in different ways. Look at the pictures of some of these groups of people shown below. Match up the pictures with the statements below numbered **1** to **8**.

For example:
Young married couple = statement **2**

How groups may be affected

1 I might have to close as most people have cars to shop in town.

2 We are just married and cannot afford an expensive house.

3 The extra noise frightens away the wildlife.

4 To get customers I have to provide food for townspeople. Villagers only want a drink.

5 I made money by selling my land so that houses could be built. Now people walk on the land I still own.

6 With all the new houses I have plenty of work to do.

7 I have to travel 10 km to school. At night there is nothing to do.

8 I came here for peace and quiet. Now I cannot drive into town and there are no buses.

Farmer | Shopkeeper | Bird-watcher | Teenager | Young married couple | Restaurant owner | Elderly person | Builder

E X T R A S

1 Activity **2** asked you to find the differences in the village between the 1890s and the 1990s. Why do you think changes have been made in:
   a) the number and type of houses
   b) the use of buildings around the green
   c) the use of the land around the village
   d) the roads?

2 It has been suggested that the woods should be cleared so that an estate of expensive houses can be built.
   a) Which groups of people will like this change?
   b) Which groups of people will be against this change?
   Suggest reasons for your answers.

Summary

Settlements change over a period of time. These changes can affect:

- the size and shape of the settlement
- the environment, e.g. new roads, larger villages
- the lives of people living in the settlement.

Why are there different land use patterns in towns?

We have seen (page 52) that when each town first began to grow it had one particular use or function. Towns and cities of today often have several different functions. The main functions are **commerce** (shops and offices), **industry** (factories), **residential** (flats and houses) and **open space** (parks). As each function tends to be found in a particular part of a town, then a pattern of land use develops. Although no two towns will have exactly the same pattern of land use, most have similar patterns. When a very simple map is drawn to show these similarities in land use it is called an **urban model**. A model is when a real situation is made simple so that it is easier to understand.

The model in diagram **A** shows a line (**transect**) from a city centre to the city boundary. It shows four main types of land use. It is suggested that this pattern developed because:

1 The oldest part of a town was in the middle. As the town grew, larger new buildings were built on its edges.
2 Land in the city centre is expensive to buy. This is because many different land users would like this site and so compete for it. Usually the price of land falls towards the edges of a town.

As a result, differences in land use can be shown by a series of **circles** drawn around the city centre (diagram **B**).

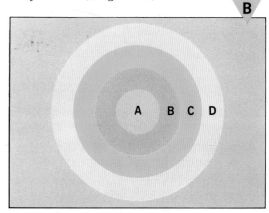

B

A   B C   D

A

Zone A

The centre of the town was the first place to be built. It is full of shops, offices, banks and restaurants. There are very few houses here.

It is now called the central business district (**CBD** for short). Most cities have been modernised.

Zone B

This zone used to be full of large factories and rows of terraced housing built in the last century. Houses were small because land was expensive.

This zone is called the **inner city**. Today many of the big factories have closed and the oldest houses have been modernised or replaced.

Activities

1 Copy out the following sentences using the correct word from each pair in brackets.
- The centre of a town is called the (ABC/CBD).
- It has many (large/small) shops and tall blocks of (flats/offices).
- There are (a lot of/few) houses and (much/little) open space.

E X T R A

2 a) Copy diagram **B** (page 62). Colour in the four areas. Give the diagram a title.
  b) Add to the drawing the name of each area – *CBD, inner city, inner suburbs* and *outer suburbs*.
  c) Name **one** part (district) of your local town that fits each area.

Using your local newspaper, try to find adverts of houses for sale in **Zones B, C** and **D**. How does this compare with diagram **A** and activity **3**?

3 a) Make a larger, simple copy of diagram **C** which shows differences in houses in three different parts of a town.
  b) **Zone C** has been done for you. Now complete **Zones B** and **D** using the correct words from the following list.

- detached
- old
- gardens
- garages
- large
- terraced
- new
- no gardens
- no garages
- small

Summary

Land in a town can be used in different ways. Land use depends upon the main function of that part of a town. The main function, or land use, of an area may result from its age, the cost of land and the lines of communication.

C

Zone B

Zone C
Semi-detached   Garage
Garden   Inter-war, 1930s   Medium-sized

Zone D

Zone C

This zone is nearly all houses built in the 1920s and 1930s.

This zone is called the **inner suburbs**.

Zone D

This zone has many large, modern houses as well as some council estates. Recently small, modern industries and large shops have been built here. There are often areas of open space.

Houses and shops and new industry are here because the land is cheaper. This is the **outer suburbs**.

Countryside

Why does land use in towns change?

Land use in towns changes over time. City centres are modernised to attract more people, while open space on the outskirts is turned into housing estates and large shopping centres.

Inner city areas were often built over 100 years ago. Naturally they have aged in that time. Houses became too old and cramped to live in and factories closed down. The inner cities had to change. Photos **A** and **B** show an inner city area in London in the early part of the twentieth century.

A

London's docks in 1926. London used to be the world's biggest port. In 1965 it still employed 28,000 dockers. There was plenty of work even if it was hard, dirty and poorly paid. Another 90,000 people worked in repairing ships, in transport and in factories using goods unloaded in the port.

B

Houses were small and packed closely together. They did not have indoor toilets, hot water or bathrooms. Yet they were cheap enough for poorly paid workers to afford and created a strong 'Eastenders' community spirit.

C

London's Docklands

In 1981 the London Dockland Development Corporation (LDDC) was set up. It began by clearing the old docks and houses. Old warehouses were turned into expensive flats. Old industries were replaced by those using high technology, such as newspapers, and by the office blocks of financial firms. The city airport and Docklands Light Railway were built. The environment is being improved and the use of the land is changing. Photo **C** shows part of this area in 2000.

How does this affect people?

Between 1967 and 1981 all of London's docks closed. Goods arriving in the port now came in big boxes called **containers**. As they were loaded and unloaded by cranes the dockers were not needed. The new container ships were too big to sail up the Thames to London. As the docks became derelict many dockers, factory workers and their families moved away to find work and better quality housing.

Diagram **D** shows how the recent changes in this area have affected different groups of people. Some groups have lived here all their lives, other groups have only just moved in.

D

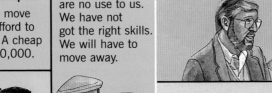

Losers

Winners

Elderly people
Shopping is expensive. Money is spent on houses and offices, not on hospitals and old people.

The LDDC
We have changed the face of London. We have created 20,000 new homes and 10,000 new jobs. The environment is becoming cleaner.

Local shopkeepers
All these newcomers mean more trade – especially as they are wealthy with money to spend.

Young married couples
We will have to move as we cannot afford to buy a new flat. A cheap flat is over £100,000.

Former dockers
The new jobs are no use to us. We have not got the right skills. We will have to move away.

School leavers
We are being trained to use computers. We should be able to get a job locally and so we can stay in the area.

Social workers
The community atmosphere has gone. The 'yuppies' who buy the luxury flats do not mix with local people.

Financial company
We moved here because of cheap land. It only takes ten minutes to travel into central London. There is high quality housing.

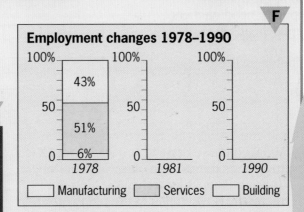

Activities

1 Table **E** below shows changes in jobs in the London Docklands between 1978 and 1990. Plot this information as a bar chart. 1978 has been done for you on diagram **F**.

E

	1978	1981	1990
Manufacturing	43%	41%	21%
Services	51%	52%	73%
Building	6%	7%	6%

F

Employment changes 1978–1990

100%

43%

51%

6%

1978

100%

1981

100%

1990

☐ Manufacturing   ☐ Services   ☐ Building

2 Divide your class into eight groups. Each group represents one of the groups named in diagram **D**.
   a) Decide if your group is a winner or a loser.
   b) Decide why your group is a winner or a loser.
   c) Choose one person to speak for your group.
   d) If your group is a **winner** say which changes you think have improved the area.
   If your group is a **loser** say what changes you would like to see take place.

Summary

As time passes, the functions and land uses of different parts of a town will change. These changes affect different groups of people in different ways.

What is a settlement hierarchy?

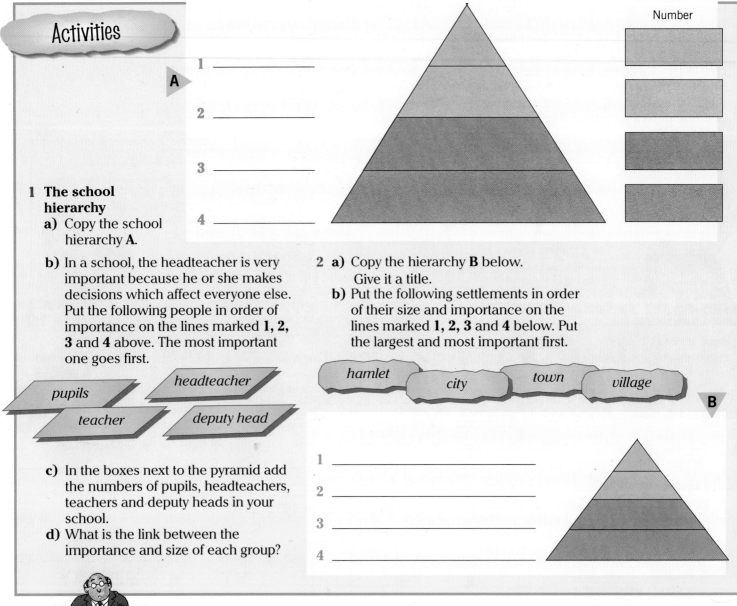

Activities

A

1 _____

2 _____

3 _____

4 _____

Number

1 The school hierarchy

a) Copy the school hierarchy **A**.

b) In a school, the headteacher is very important because he or she makes decisions which affect everyone else. Put the following people in order of importance on the lines marked **1, 2, 3** and **4** above. The most important one goes first.

pupils

headteacher

teacher

deputy head

c) In the boxes next to the pyramid add the numbers of pupils, headteachers, teachers and deputy heads in your school.

d) What is the link between the importance and size of each group?

2 a) Copy the hierarchy **B** below. Give it a title.

b) Put the following settlements in order of their size and importance on the lines marked **1, 2, 3** and **4** below. Put the largest and most important first.

hamlet   *city*   *town*   *village*

B

1 _____

2 _____

3 _____

4 _____

What you have done in activities **1** and **2** is to arrange things in an order of importance. This is called a **hierarchy**. A settlement hierarchy is when places are put in order.

It is possible to use three different methods to get a settlement hierarchy.

The three methods can be tested by making a simple statement about each one. These three statements are given in the following activities section.

1 Count the number of each type of settlement in an area, and look at their distance from each other.

2 Work out the size of each settlement. This could either be the size of the area (land) it covers or the size of its population (the number of people living in it).

3 Find out the range and number of functions and/or services it provides. A service is something which helps people, e.g. health centre, buses.

Activities

1 The larger the settlement the further it is away from another large settlement.
Look at map **C** and write down the average (**mean**) distance between the
- villages
- towns
- cities on the map.

2 The larger the settlement the fewer there are of them.
a) Look at map **C** and count the number of
- villages
- towns
- cities on the map.

b) Look at table **D**. Column **1** names five types of settlement. Column **2** suggests the minimum number of people living in each settlement. Copy out the table, matching up the two columns.

3 The larger the settlement the more services it will have.
Table **E** shows a settlement hierarchy based on services.
a) Use the Ordnance Survey (OS) map and its symbols on pages 124 and 125 to complete a copy of table **F** below.
b) Draw the correct Ordnance Survey symbol in the space below each column heading.
c) When you have completed the table, put the five places in order of importance. The one with the most services goes first.

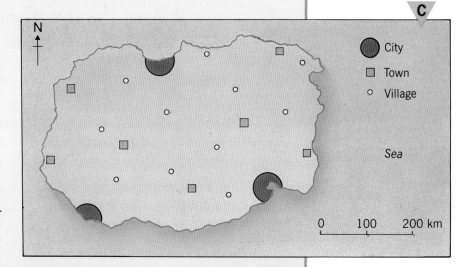

C

City · Town · Village · Sea

0   100   200 km

D

Column 1	Column 2
City	200
Town	5
Village	100,000
Hamlet	10,000
Farm	25

E

Settlement	Services
Hamlet	Perhaps none
Village	Church, post office, public house, shop for daily goods, small junior school
Town	Several shops, churches and senior school, bus station, supermarket, doctor, dentist, banks, small hospital and football team
City	Large railway station, large shopping complex, cathedral, opticians and jewellers, large hospital and football team, museum

F

Settlement	Post offices	Churches	Main roads	Telephones	Inns, public houses	Railway/ bus station	TOTAL
Foxton							
Central Cambridge							
Haslingfield							
Grantchester							
Harston							

Summary

Settlements found in any given area will vary in size and function. These settlements can be arranged in order of importance to give a hierarchy.

Where do we shop?

We all go shopping. We need to buy things so that we can feed ourselves and live our lives. Some of us shop simply for enjoyment. Recent surveys have found that shopping is one of Britain's most popular leisure activities.

Shopping is also big business. Each year, people in the UK spend over £8 billion on food shopping alone. A further £3 billion is spent on clothes and shoes. More than 2 million people in the UK work in shops, many of them part-time.

As we have seen, smaller settlements usually have very few shops. Larger settlements, however, are likely to have several shopping centres. In recent years, many of these have been built out of town and well away from the city centre.

Where people choose to shop depends upon what they want to buy and how often they need that product. The larger the shopping centre the more choice of goods and services there will be. People will travel long distances to these centres.

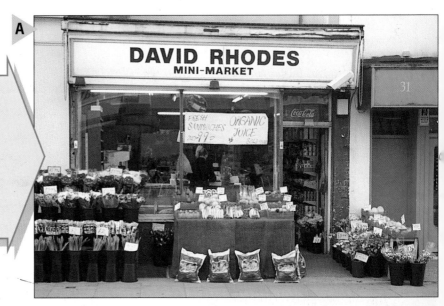

A

- Some **goods** such as food and newspapers don't cost very much and may be needed every day. We are happy to buy them in the nearest convenient place.
- These are called **convenience** or **low order goods**. They may be bought at the local **corner shop**, nearby shopping centre or supermarket.

B

- Some goods like clothes and furniture are much more expensive. We buy them less often and like to compare styles and prices before we buy.
- These are called **comparison** or **high order goods**. They are bought at large shopping centres, either in the city centre or out of town. Here, there is usually a good choice and lower prices because of competition.

Activities

1 Complete these sentences.
  a) Convenience goods are ...
  b) Comparison goods are ...
  c) A corner shop is ...

2 Look at the goods shown in drawing **C**. Sort them into two groups: convenience goods and comparison goods.

3 Now decide where you would buy the goods in drawing **C**. Look at drawing **D**, which gives you three choices. You must make a separate trip to each centre. Write out a shopping list for each visit.

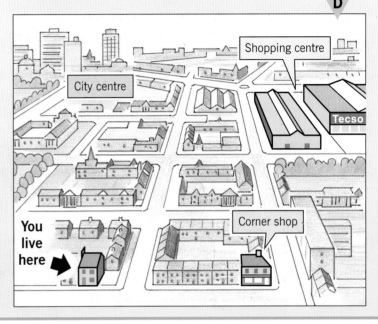

E X T R A

You may be close enough to a shopping centre to visit it during a geography lesson. If you do you **must** go in groups and take great care when crossing roads.

Aim – to compare the shopping habits of people at your local shopping centre with those of people in the city centre.

Equipment – questionnaire, clipboard and pencil.

Method
a) Make up a questionnaire similar to the completed one on the right.
b) Politely ask at least 20 male and female shoppers of different ages the questions.
c) Share your answers with other groups.
d) Think of ways to illustrate your results.
e) Describe your findings.
f) Suggest reasons for differences between the results of your questionnaire and those of the one taken in the city centre.

Shopping survey

The numbers on the right show the result of asking 100 people in a city centre shopping area (mall) the following questions:

1 Do you shop here
  ● every day? — 15
  ● two or three times a week? — 15
  ● once a week? — 50
  ● once a month? — 20

2 Have you travelled
  ● less than 1 mile? — 15
  ● between 1 and 2 miles? — 20
  ● between 2 and 5 miles? — 30
  ● over 5 miles? — 35

3 Do you travel here
  ● on foot? — 5
  ● by car? — 75
  ● by bus? — 15
  ● any other way? — 5

4 Do you do most of your weekly shopping here?
  ● Yes — 75
  ● No — 25

5 What is the main thing you buy here?
  ● Food — 30
  ● Clothes — 40
  ● Furniture — 10
  ● Domestic equipment — 10
  ● Others — 10

Summary

There are many different types of shopping centre. The larger the centre, the greater the choice of shops and goods there are to buy.

How has shopping changed?

The city centre is the main shopping area in a town. It has the largest number of shops, the biggest shops and the most shoppers. People are willing to travel long distances to the city centre because of the great choice of goods that they can buy there.

The main advantage of the city centre is its **accessibility**. Most of the main roads, bus routes and rail systems from the suburbs and surrounding areas meet at the city centre. It is therefore the easiest place for most people living in the town to reach.

City centre shopping has changed a lot recently. Attempts have been made to reduce traffic congestion, and to provide for the safety and comfort of shoppers. Many towns now have covered **shopping malls** which give protection from the weather and are traffic free.

A

- Almost anything can be bought here. There are large department stores, nationwide supermarkets, chain stores and many specialist shops. Competition between shops keeps prices low.
- Good accessibility by car and public transport make it the most visited and busiest type of shopping centre.
- Overcrowding and traffic congestion cause problems. Pedestrianised areas and covered walkways have improved shopper comfort.

City centre

Activities

1 **a)** Name the main street, shopping area or mall in your local town or city.
   b) Name one department store, one nationwide supermarket and three specialist shops found in it.

2 Make a larger copy of sketch **B** showing a city centre shopping centre. Complete it by answering the questions.

B

a) What makes the city centre the main shopping centre?

b) Why are people willing to travel long distances to this centre?

c) Name four different types of shops in the city centre.

d) What three things make the city centre accessible?

e) What are the main problems of these centres?

f) What is being done to improve city centre shopping?

The biggest change in shopping has been the development of huge out-of-town shopping centres. These are located on the edge of cities, usually next to a main road or motorway. They are designed to attract motorists from a wide area and offer good accessibility and free parking.

Many people are worried about the development of these centres. They are concerned that they take trade away from the traditional shopping outlets. Some city centres have lost up to half of their business in the last ten years and are in serious decline. Many smaller shopping parades and corner shops have had to close altogether.

Another drawback of out-of-town shopping is the increased use of cars. This has caused more air pollution, noise and traffic congestion in suburban areas.

The government is very concerned about these effects. In future, it might not give permission to build any more out-of-town centres.

C

- Ideal for shopping by car, with good road access and free car parking.

- Contain a large number of shops with a wide choice of goods. Prices are kept low by bulk buying and low running costs.

- Popular with shoppers who enjoy the bright and attractive air-conditioned shopping malls. Security staff ensure safety for families.

- Most centres have cafés, restaurants, cinemas and a wide range of other leisure amenities.

Out of town

D

Visit Merrydale for better shopping

3 Design a poster to show the advantages of using a shopping centre like the one shown in photo **C**. Your poster should be attractive and interesting, and show the facts.

4 Imagine that you own a small shop close to a new out-of-town shopping centre. Your profits are down and you think you may soon have to close. Write a letter to the local council to say that the centre was a mistake. Mention all the ways that you think it is harming the area.

Summary

Shopping habits are changing. The city centre has always been the main shopping area in a town but it is now often congested and expensive. As more people shop by car, modern out-of-town centres are becoming increasingly popular.

The settlement enquiry

This enquiry is linked to the Settlement section on pages 52 to 75 of this book. You may need to refer to those pages, particularly when you are working on your conclusion. The Ordnance Survey (OS) key and map on pages 124 and 125 will also be useful.

Your task is to answer the enquiry question given below. To do this you will need to use information about four small settlements in the Cambridge area. You could work by yourself, with a partner or in a small group. Your finished enquiry could be in the form of a booklet or a display to be put on the classroom wall. You might be able to use a computer to word process your work and help with your graphs.

> ## How does the size of a settlement affect the number of goods and services it provides?

1 Introduction – what is the enquiry about?

a) Look carefully at the enquiry question and say what you are going to try to find out. Remember to mention that you will use settlements in the Cambridge area to do that. Explain what goods and services are. Pages 66, 67, 74 and the Glossary will help you.

b) Describe the area using maps, labelled sketches and writing.

First I shall need to explain what the enquiry is about.

Now how could I answer this?

Next I shall have to decide what information I need and where I could get it from.

Then I could collect the information and present it in an interesting and clear way.

Finally I could describe my findings and try to suggest reasons for them.

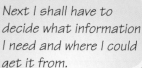
Description

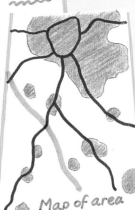

Location of Cambridge

Sketch map of shops at Foxton

Map of area

The public house at the junction in Newton

Newton

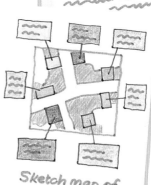

Harston

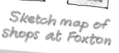
Harston post office is on the busy A10 road

2 How large are the settlements?

a) By **population** – figures are given with the village plans on pages 74 and 75. Draw a bar graph to show your results. Arrange the bars in order of size with the largest on the left.

b) By **area** – measure this using the OS map on page 125.

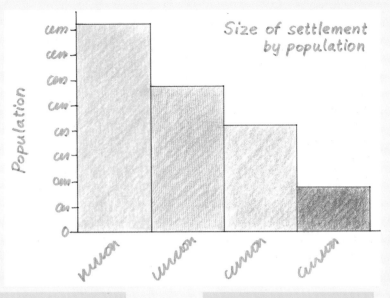

Size of settlement by population

Population

| Put a tracing paper grid over the map. | Count the squares covering each village. | Graph your results. |

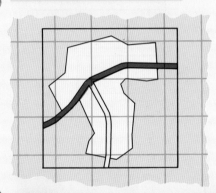

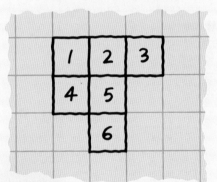

Size of settlement by area

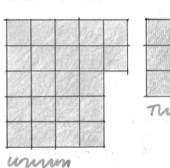

Barton

'Conkers' stores and post office in Barton

The post office in the village of Foxton

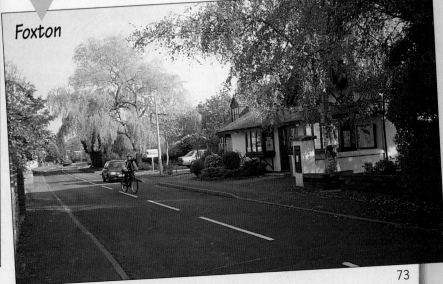

Foxton

3 How many goods and services are there?

Use the village plans and the OS map to find this out for each settlement.

- Count the number of **different** goods available, not just the total number of shops.
- If there is more than one of the same item in a village, give the **total** number. For example, two garages would count as two different items.

Key Geography Encyclopedia

File   Edit   Services   Mail   Special   Window   Help

Goods are products that are useful to people, e.g. food, clothing, jewellery, stationery, newspapers, toiletries.

Services are things that help people, e.g. post offices, churches, telephones, railway stations, garages, main roads. Services do not provide a product.

Count the goods and services for each village. ➡ Write up your results in a table. ➡ Graph your results with the highest number at the top.

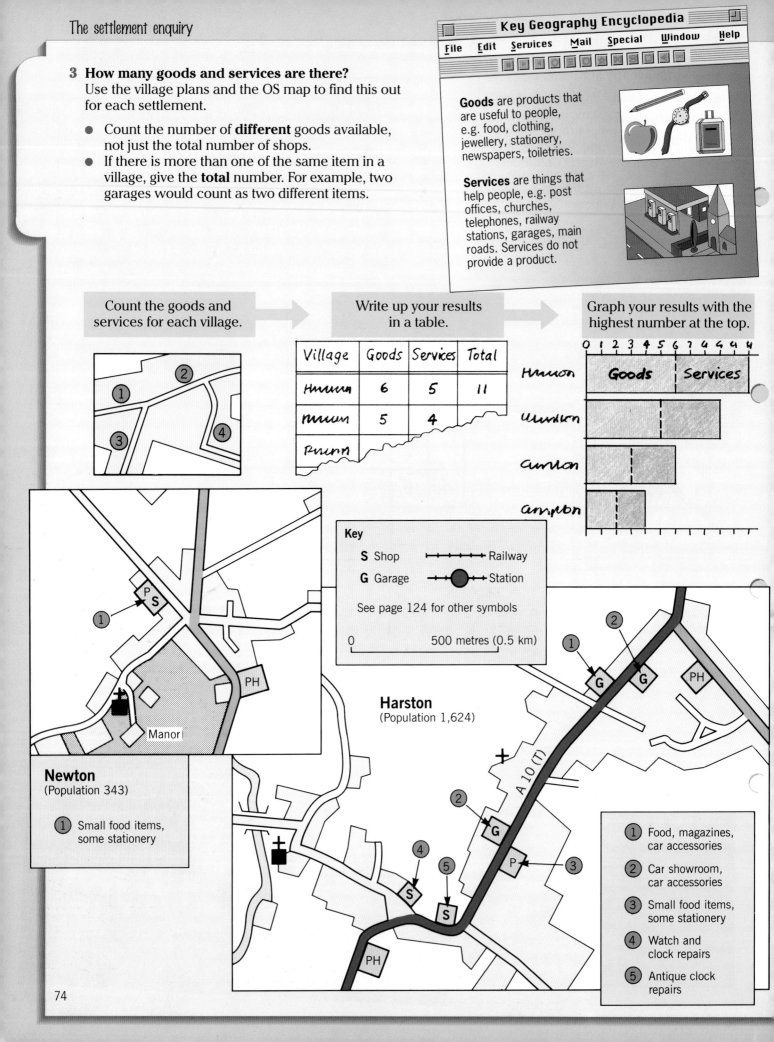

Village	Goods	Services	Total
Harston	6	5	11
Newton	5	4	
Burwn			

Key

S Shop   ├─┼─┼─┤ Railway
G Garage ├─●─┤ Station

See page 124 for other symbols

0 ———— 500 metres (0.5 km)

Newton
(Population 343)

① Small food items, some stationery

Harston
(Population 1,624)

① Food, magazines, car accessories

② Car showroom, car accessories

③ Small food items, some stationery

④ Watch and clock repairs

⑤ Antique clock repairs

4 Conclusion

Now you must look carefully at your work and answer the enquiry question. Notice that it begins 'How . . .'. That means you will need to both describe your findings **and** try to explain them.

a) Make a summary of what you have found out. You could draw up a simple table and do some writing for this. Make sure that you keep to the question asked.

b) Suggest reasons for your findings.

Settlements in order ...	
... of size (largest first)	... of number of goods and services (most first)
1	1
2	2
3	3

Description
This enquiry has shown that ...

For example ...

Now this should be easy and straightforward.

Now are there any other reasons I can think of?

Goods and services are for people.

How will main roads affect the need for goods and services? I wonder why there are so many garages?

Surely the more people there are, the more goods and services there will be.

Not all people will be local, though. What about those who come from somewhere else?

So which settlements have the most people in them?

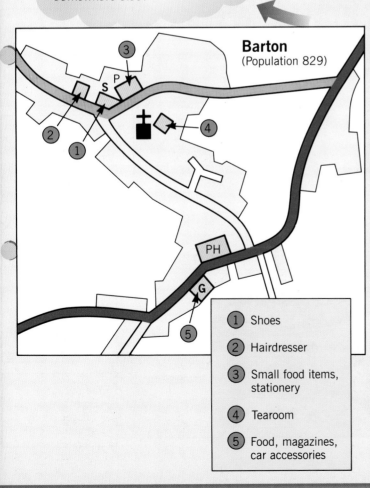

Barton
(Population 829)

1. Shoes
2. Hairdresser
3. Small food items, stationery
4. Tearoom
5. Food, magazines, car accessories

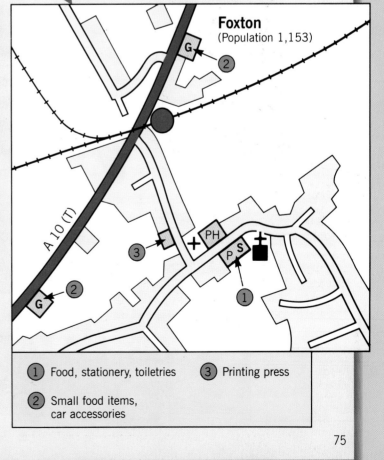

Foxton
(Population 1,153)

A 10 (T)

1. Food, stationery, toiletries
2. Small food items, car accessories
3. Printing press

5 Transport

Which transport to use?

Transport is used to carry people and goods from one place to another. People need transport to go to work, to travel to the shops, to go on holiday and to visit friends. When people travel they are called passengers. When goods are moved from one place to another they are called freight or cargo. Coal, wheat and televisions are examples of goods that are often carried as freight or cargo.

The type of transport used depends on many things. The most important ones are distance, time, cost and the size of the things being carried. These points, and others, are shown in diagram A.

There are many different types of transport. Some are fast and some are slow. Some are cheap and some are expensive. Some are better for cargo than for passengers. Some serve only a few places whilst others can go almost anywhere. Diagrams B, C, D and E show some of the features of road, rail, sea and air transport.

A

Which transport?

Time · Cost · Distance · Value of goods · Size of goods · Weight of goods

B

Road transport

Includes:
- Cars
- Lorries
- Buses
- Motorbikes

Needs:
- Large amounts of land

Easy travel from place to place

Choose who to travel with

Cheaper than rail or air

Travel when you want to

Good for:
- Passengers
- Private use
- Freight
- Short distances
- Door to door journeys

Bad because:
- Slow
- Dangerous
- Exhaust fumes
- Traffic jams

C

Rail transport

Includes:
- Passenger trains
- Freight trains

Needs:
- Track and stations

Comfortable

Reliable

Cheaper than air

Can work while travelling

Fast and safe

Good for:
- Medium length journeys
- Passengers
- Heavy and bulky goods
- Intercity use

Bad because:
- Expensive
- Villages poorly served

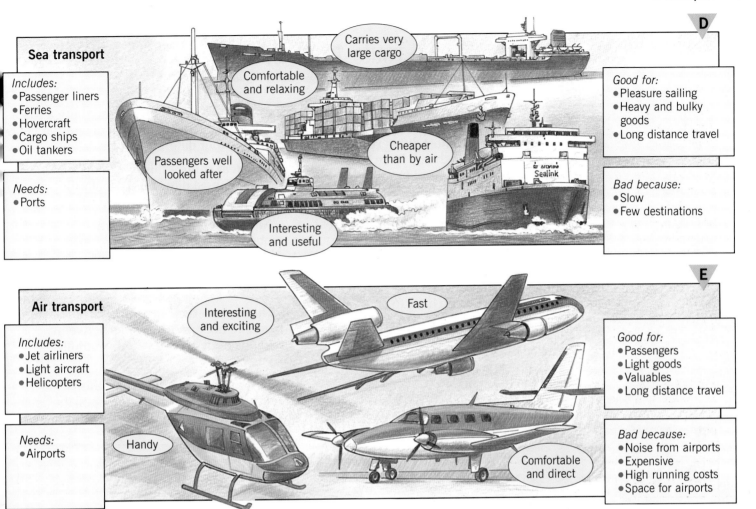

Sea transport

Includes:
- Passenger liners
- Ferries
- Hovercraft
- Cargo ships
- Oil tankers

Needs:
- Ports

Carries very large cargo

Comfortable and relaxing

Passengers well looked after

Cheaper than by air

Interesting and useful

Good for:
- Pleasure sailing
- Heavy and bulky goods
- Long distance travel

Bad because:
- Slow
- Few destinations

Air transport

Includes:
- Jet airliners
- Light aircraft
- Helicopters

Needs:
- Airports

Interesting and exciting

Fast

Handy

Comfortable and direct

Good for:
- Passengers
- Light goods
- Valuables
- Long distance travel

Bad because:
- Noise from airports
- Expensive
- High running costs
- Space for airports

Activities

1 **a)** How do you usually travel to school?
  b) Which things from diagram **A** help you to choose that method of travel?

2 Give the meaning of each of the following:

transport

passengers

freight and cargo

3 **a)** Make a copy of table **F** below.
  b) Match the best type of transport with each journey.
    Tick only **one** box for each journey.
  c) Choose any **two** journeys and explain your choice.

F

Journey	Road	Rail	Air	Sea
Short holiday to Spain				
Important letter to America				
Business trip to London				
Take coal to Japan				
Visit friends 10 km away				

Summary

There are many different types of transport. They all have advantages and disadvantages. When choosing which transport to use for a journey, it is important to think about distance, time, cost and what is to be carried.

Which route to take?

A straight line is the shortest distance between two places. Some roads and railways follow this most direct **route** and it is usually quick and cheap. More often a route will make twists and turns as it goes from one place to another. These are called **detours**. What are the reasons for this? Why don't all routes travel in straight lines?

Think about how you get to school. Almost certainly you will not travel in a straight line. Perhaps there are buildings or private land in the way. Maybe you have to cross a road or use a bridge to take you over a railway or stream. Meeting up with friends may also take you out of the way. Together, all these things affect which route you take from your home to school.

Some of the things affecting the route a road or railway takes are shown in diagram **A**. There are four different factors:

1 **Shortest route** – this is a straight line. It is the most direct and usually the quickest and cheapest way.

2 **Natural features** – these are obstructions like mountains, rivers, marshy land and coastal inlets. They may cause routes to detour from a straight line.

3 **Human needs** – these are things that people need, such as links between towns or by-passes to reduce traffic jams.

4 **Environmental concerns** – this is where the route taken may spoil the surroundings.

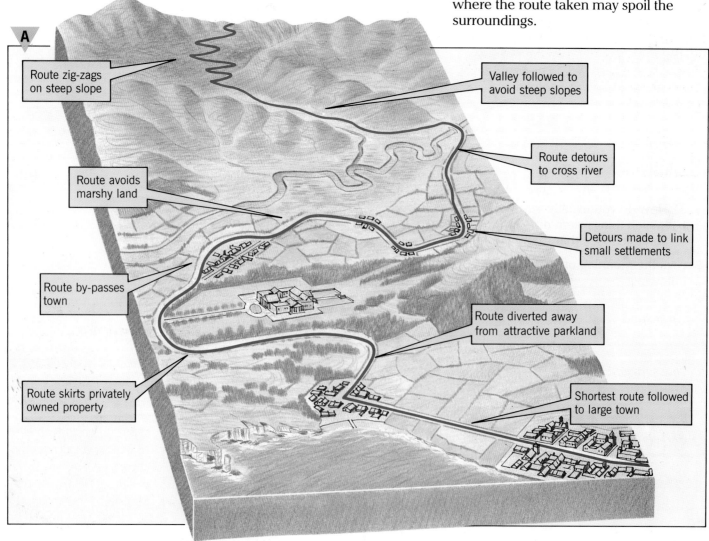

A

Route zig-zags on steep slope

Valley followed to avoid steep slopes

Route detours to cross river

Route avoids marshy land

Detours made to link small settlements

Route by-passes town

Route diverted away from attractive parkland

Route skirts privately owned property

Shortest route followed to large town

B Railway across the Nullarbor Plain, Australia

C Road over the Andes Mountains, Peru, in South America

Photos **B** and **C** show two very different routes. The railway in Australia (photo **B**) is the longest straight section of track in the world and stretches an incredible 478 kilometres without a single bend. It is straight because there are no natural obstructions to avoid, no settlements to divert to, no worries about damaging the environment. It is a perfect example of a route that takes the shortest distance between two points. The road in Peru (photo **C**) is very different. It follows a valley but has to zig-zag to go up the sides of it, otherwise the road would be too steep. This is an example of a route affected by natural features.

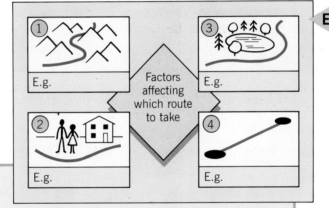

E

Activities

1 Look at map **D** which shows the route that Claire takes to her school.
   a) Describe her route. Start with: *Claire leaves her house and turns right…*
   b) Give **three** reasons why she doesn't travel in a straight line from her home to the school.

2 Make a larger copy of all of diagram **E**.
   a) Give each box a suitable title.
   b) Add an example to each one.

3 Look at diagram **A** and list the things that affect the choice of route. Give your answers under these headings. The first one has been done for you.

> **Shortest route** – is a straight line between places. It is usually the quickest and cheapest.
> **Natural features** –
> **Human needs** –
> **Environmental concerns** –

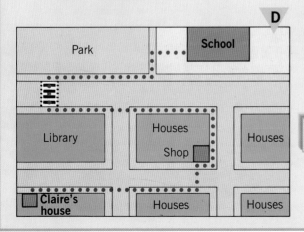

D

Park | School
Library
Houses | Shop | Houses
Claire's house | Houses | Houses

E X T R A S

1a) Draw a simple map to show your route to school.

b) List the reasons why you do not travel in a straight line.

Summary

The route a road or railway takes depends on physical, human and environmental factors. A straight line is the shortest route but it is often not possible to follow that route.

Developments in transport — the good news . . .

Changes in transport have been rapid and spectacular. Journeys that were once expeditions now happen every day. Places that were visited only by explorers are now seen by tourists. In the 1990s no place in the world is more than 24 hours away. The world is not shrinking. Distances still remain the same. What has happened is that improvements in types of transport and in transport networks have made places easier to get to.

A place that is easy to get to is said to be **accessible**. Accessibility depends on distance, the number of routes and links, travel is faster and the actual cost of a journey has decreased.

Better accessibility can bring many benefits. These include less time travelling, cheaper travel, a greater choice of holiday destinations, more markets for industrial products and increased trade.

A

French supertrain sets a world speed record of 270 km/h (168 m.p.h.)

Concorde reaches speed of 2,333 km/h (1,450 m.p.h.) and crosses Atlantic in 2 hrs 55 mins

Luxury coaches give more comfort at less cost

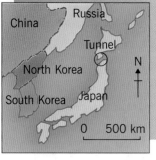

Japan's northern outpost linked to the main island by the world's longest rail tunnel

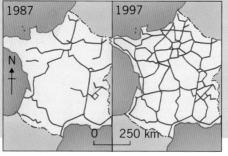

French plan massive growth in motorway system

Channel Tunnel halves crossing time

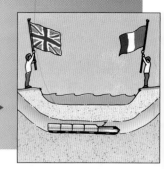

Activities

1 **a)** Copy and complete the accessibility crossword by adding five things that affect accessibility. *Clue:* the answers are all given in one sentence on this page!
   b) Give the meaning of accessibility.

2 From **A** above, give examples of transport improvements which will:
   a) reduce travelling times
   b) produce a better network.

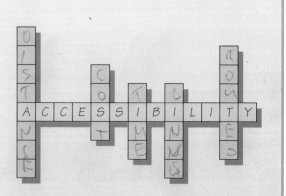

3 Describe an example of improved transport in the area where you live.

... and the bad news

Unfortunately, although developments in transport have brought many benefits, they have also brought problems. Sometimes transport is just too popular and the networks cannot cope. Roads get jammed, car parks fill up, trains are packed and airports burst at the seams.

Another more serious problem is the damage that transport can do to the environment. This is called **environmental pollution**. Pollution is dirt or noise or anything else that is damaging to an area. It may be harmful to people, to animals, to plants or to the surroundings as a whole. All forms of transport damage the environment in some way, but we need transport and transport improvements. So what we must do is to be very careful to balance the advantages of progress against the disadvantages of spoiling the world in which we live. We must try to increase the good news but reduce the bad news.

B

NEWTOWN JOURNAL
City fumes over car exhaust pollution

Morning News
Famous view spoilt by new car park

Evening Herald
New road threat to Amazon rain forest

THE CHRONICLE
Residents to be heard over runway noise problem

The Oracle
Long delays expected as airport congestion gets worse

THE DAILY GLOBE
Multiple crash closes over-used motorway

The Post
OIL SLICK REACHES BEACHES AS DOOMED TANKER SINKS

The Daily Times
Vineyards lost as French supertrain route gets OK

Weekender
10 mile traffic jam hits M25

WEEKLY LIFE
DELAYS AGAIN AS ROADWORKS RETURN

Activities

1 **a)** Unjumble the following words to give four types of pollution.

IAR	DNAL	TRAWE	SNIOE

   b) Give the meaning of the term **environmental pollution**.

2 Make a larger copy of diagram **C** and complete it using information from this page.

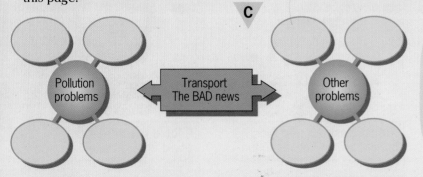

C

Pollution problems — Transport The BAD news — Other problems

E X T R A

Imagine that a motorway is to be built very near to your house.
- List the advantages and the disadvantages that the motorway would bring.
- Would you be in favour of, or against, this development? Give reasons for your answer.

Summary

Developments in transport are happening all the time. An important advantage is improved accessibility. One major disadvantage is environmental pollution. Care must be taken to protect the environment from damaging transport developments.

Traffic in urban areas — why is it a problem?

Traffic is a serious problem in most urban areas. Large cities such as London, Paris and New York have over a million cars trying to move around in their central areas. Movement is often impossible. Perhaps worse than congestion is the problem of pollution. Exhaust fumes are poisonous and can seriously damage health. Some city workers are so concerned that they wear masks to protect themselves. So what can be done? What are the causes of the problem, and why have they not been solved?

What is the problem?

Look at any urban area and you will soon be able to answer this question. Cars, buses and lorries all over the place cause congestion and chaos. They produce a lot of fumes, noise and danger. Other effects are:

- traffic jams blocking roads and stopping all movement
- delays for police, fire service and ambulances
- slow movement of people and goods
- loss of business and money
- people and buildings affected by noise and vibrations
- danger from accidents
- harmful exhaust fumes
- lack of parking places.

We live in an age of rapid transport yet vehicle movement is now actually slower than it was 80 years ago.

A

P462 PLP

Activities

1 Look at the information with photo **A**. List what you think are the five worst problems caused by increased traffic in towns.

2 The two people in drawing **B** are badly affected by traffic congestion and pollution. For each person write a letter to the local MP explaining how the problem affects them personally.

B

A businessman who lives out of town and drives to the city centre each day to work

A local resident with two young children who lives close to the main road

What is the cause?

There are many reasons for the traffic problems in our cities. The main one is simply that the number of cars has increased at a tremendous rate and there are now too many cars for cities to handle. It is predicted that this increase in cars will continue. By 2025 the number of cars might double and the number of lorries be three times greater than in 2000.

Another reason for traffic problems in cities is that most city centres were designed and built before cars were invented. They are therefore just not suited to today's transport. The problem is worst in the morning and in the late afternoon when people are travelling to and from work. This is called **the rush hour**.

C

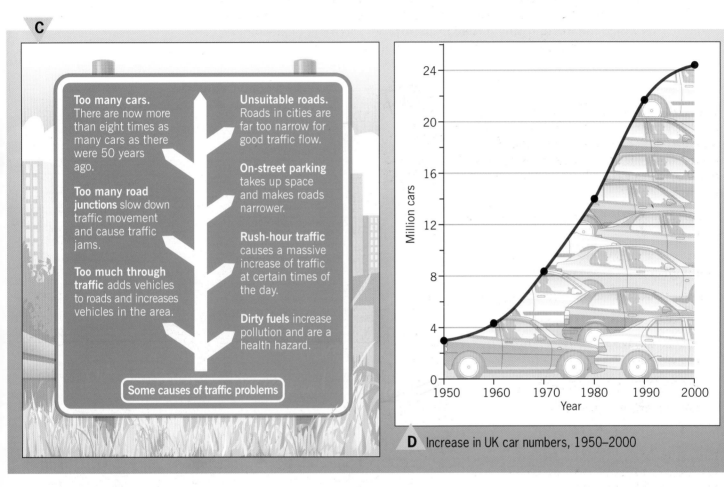

Too many cars. There are now more than eight times as many cars as there were 50 years ago.

Too many road junctions slow down traffic movement and cause traffic jams.

Too much through traffic adds vehicles to roads and increases vehicles in the area.

Unsuitable roads. Roads in cities are far too narrow for good traffic flow.

On-street parking takes up space and makes roads narrower.

Rush-hour traffic causes a massive increase of traffic at certain times of the day.

Dirty fuels increase pollution and are a health hazard.

Some causes of traffic problems

D Increase in UK car numbers, 1950–2000

E Graph descriptions

3 Look at drawing **C**. List what you think are the five worst problems caused by increased traffic in cities. You need only write out the words in **bold**.

4 Look at graph **D**.
   a) How many cars were on Britain's roads in 1950? How many were there in 2000?
   b) Which of the graphs in **E** looks most like graph **D**? Use that description to describe the change in car numbers between 1950 and 2000.

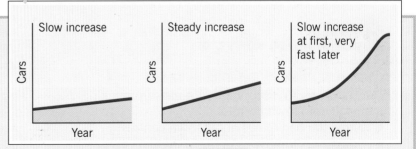

Summary Congestion and pollution are major problems in urban areas. The main causes of these problems are too many cars, rush-hour traffic and unsuitable roads.

Traffic in urban areas — is there a solution?

There are two main ways of approaching the problem. The first is to allow **private transport** to increase and make improvements to cope with larger amounts of traffic. The second is to restrict private transport and discourage motorists from bringing cars into town centres. This would mean improving **public transport** such as bus and train services.

In fact, the traffic problem is so big and complicated that no single solution will ever completely solve it. The best way is to try to reduce the worst parts of the problem by using several solutions together. Some ideas that have been tried are shown in diagram **A**. Can you think of any others?

A

Encourage private transport

Better traffic management:
- roundabouts
- one-way systems
- traffic lights

More off-street parking

Urban motorways to improve traffic flow

By-passes to keep through traffic out of towns

Discourage private transport

Improve public transport:
- reduce fares
- increase speed and comfort
- develop rail routes like the London underground

Park-and-ride schemes:
- leave cars on town outskirts
- travel by free bus to centre

Make car parking difficult:
- increase charges
- reduce spaces

Activities

1 a) What is meant by public transport?
   b) What is meant by private transport?
   c) Name each of the following types of transport and sort them into **Public** and **Private**.

2 Draw a poster to discourage motorists from taking their cars into town centres.
   - Show the bad things about town centres.
   - Show the other types of transport that can be used.
   - Colour your poster and make it interesting and attractive.

3 Diagram **B** shows how private and public transport in towns affect each other.
   Copy and complete the diagram using the following phrases.

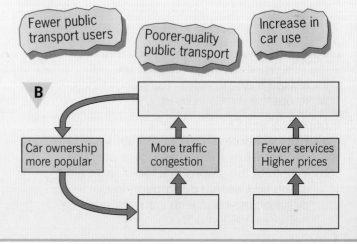

Fewer public transport users

Poorer-quality public transport

Increase in car use

B

Car ownership more popular

More traffic congestion

Fewer services Higher prices

Public transport systems

In an attempt to reduce congestion and pollution, some cities have built new public transport systems. The aim of these systems is to move people around as safely, quickly and cheaply as possible. They are also designed to reduce pollution and so protect the environment.

Manchester's Metrolink tram system is an example of a surface light railway. This means that it runs on rails along existing roads and sections of former railway. Priority is given to the tram at traffic lights and road crossings. Travel by Metrolink has proven faster and cheaper to use than car transport.

 C

The Manchester Metrolink

- Connects suburbs to city centre
- Provides convenient way to travel in city centre
- Links with existing bus and rail routes
- Trams every 6 minutes during daytime
- Each tram can carry 206 passengers
- Fares subsidised to provide cheap travel
- Automatic ticket machines reduce costs

- Speeds up to 80 km/hr on former railways
- 26 stations and over 29 km of track
- Excellent wheelchair and pram access
- Powered by electricity so reduces noise and air pollution
- Carries over 16 million passengers a year
- Takes up to 2.5 million car journeys a year off the road

4 How does the Metrolink:
   a) provide a fast and cheap service
   b) serve people living away from the system
   c) help protect the environment?

5 Public transport is not popular with everyone. Make a list of the disadvantages of a system such as Metrolink.

6 Describe a traffic problem near your school or where you live. Suggest how the problem could be reduced.

Summary Solving the problem of urban traffic is difficult. Better public transport may be the best way to improve people's movements without further damaging the environment.

Where should the by-pass go?

When a place becomes too crowded with vehicles (congested), a road can be built around it to take away some of the traffic. A road that is built to avoid a congested area is called a **by-pass**. Some by-passes are very long. The M25 which goes all the way round London is over 160 km (100 miles) long. Most by-passes are much shorter than this.

Building a by-pass is not easy. Money has to be found, suitable routes planned out, and discussions held between people whom the route may affect. This is all very difficult and takes a long time.

A

Considerations	Red route	Blue route	Yellow route
Is the shortest route			
Avoids all the built-up area			
Avoids best farmland			
Avoids steep slopes			
Avoids marshy areas			
Avoids beauty spot			
Avoids parkland			
Needs fewest bridges			
Requires fewest trees to be cut down			
Serves the industrial estate			
Total			

Activities

1 Look carefully at diagram **B**. It shows an imaginary place called Keytown and some of the countryside around it.

Keytown was once a pretty, quiet village with narrow streets and an attractive central square. It is now busy, congested and can no longer cope with all the traffic going through it. It has been decided to build a by-pass to reduce the traffic going through the town centre.

Three possible routes have been suggested for the by-pass. Your task is to choose the best one.
   a) Copy table **A**, which shows some things that should be considered when choosing a by-pass route.
   b) Show advantages of each route by putting ticks in the *Red, Blue* or *Yellow* columns. More than one column may be ticked for each point.
   c) Add up the ticks to find which route has the most advantages.
   d) Which route would you choose? The one with the most advantages would be best.
   e) Give **two** disadvantages of your chosen route.
   f) Describe the route you have chosen. Start with, 'The by-pass leaves the main road at…'.

2 The *Yellow route* would be a popular choice because it follows a disused railway line and no property would need to be knocked down.

Work in pairs and suggest which of the people below would be against the *Yellow route*. Give reasons for your answer.

Walkers

Managers of factories on industrial estate

Residents of new housing estate

Hotel owner in Keytown

B

Keytown – suggested by-pass routes

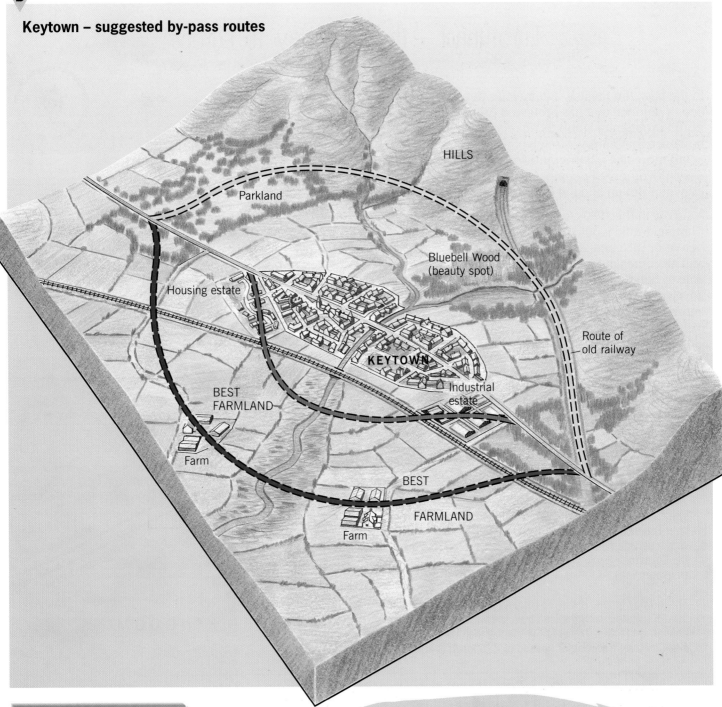

HILLS

Parkland

Bluebell Wood
(beauty spot)

Housing estate

Route of
old railway

KEYTOWN

BEST
FARMLAND

Industrial
estate

Farm

BEST

FARMLAND

Farm

E X T R A

Find out about a by-pass plan in your
local area.

- Draw a simple map to show where the
  by-pass goes.
- Briefly describe the route.
- Say why the by-pass was needed.
- Give the advantages and
  disadvantages of the route chosen.

Summary

A by-pass is one method of reducing
congestion in busy areas. Choosing
the route for a by-pass is very
difficult. Cost, the availability of land
and a concern for the environment are
important considerations. No route
will satisfy everyone.

Eurotunnel — the new way to Europe

In the past, the English Channel has been a line of defence that has protected Britain from European neighbours. In recent times, however, that narrow stretch of water has acted more like a barrier preventing the easy movement of people and goods between Britain and the Continent.

A survey in the 1980s forecast that cross-Channel traffic would more than double between 1983 and 2003. The existing sea and air routes at that time would not be able to handle that increase. In 1987 the governments of Britain and France finally agreed to build a **Channel Tunnel** to try to overcome the problem.

As diagram **A** shows, not everyone at the time was in favour of the tunnel. Eventually, however, after much delay it was completed and finally opened in May 1994. The tunnel provides a link between the transport networks of Britain and mainland Europe. It makes going to the Continent faster and easier. It also helps to improve tourism, trade, and industry.

A

For

The 1980s tunnel debate

(Many of these points are still valid today, years after the tunnel was opened.)

1 Road and rail networks of Britain and Europe joined
2 Faster travel
3 Trade and industry helped
4 Up to 35,000 new jobs created
5 Up to 1,500 lorries a day taken off the roads – this means less congestion and less pollution
6 No delays due to bad weather
7 The tunnel is a symbol of unity between Britain and Europe

The tunnel is about 50 km long and runs between terminals near Folkestone and Calais. There are two separate rail tunnels with a service tunnel between them (see sketch **D**). Fast trains – the Eurostar – carry through traffic, and special shuttle trains transport vehicles.

The journey time is about 35 minutes and trains leave every 10–15 minutes. A high speed train link will eventually connect London with Paris. Travel to Europe will then be on a fast InterCity train travelling at speeds of 225 km/h (140 m.p.h.).

LONDON

Chatham

Motorway

Maidstone

High speed train

Ashford

UNITED KINGDOM

Hastings

ENGLISH CHANNEL

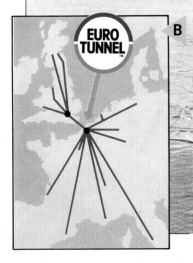

B

EURO TUNNEL

C

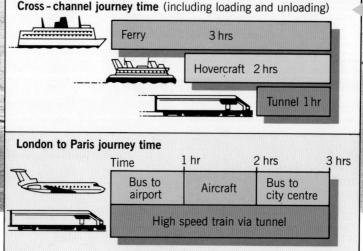

Cross-channel journey time (including loading and unloading)

Ferry 3 hrs

Hovercraft 2 hrs

Tunnel 1 hr

London to Paris journey time

	Time	1 hr	2 hrs	3 hrs
Aircraft	Bus to airport	Aircraft	Bus to city centre	
Train	High speed train via tunnel			

Summary

The Channel Tunnel brings the road and rail networks of Britain and mainland Europe together. This increases accessibility and helps improve trade and industry. Many people are worried about the damaging effects the tunnel project can have on their lives and surroundings.

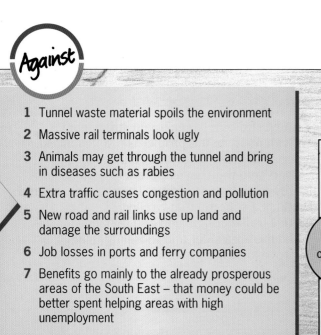

Against

1 Tunnel waste material spoils the environment

2 Massive rail terminals look ugly

3 Animals may get through the tunnel and bring in diseases such as rabies

4 Extra traffic causes congestion and pollution

5 New road and rail links use up land and damage the surroundings

6 Job losses in ports and ferry companies

7 Benefits go mainly to the already prosperous areas of the South East – that money could be better spent helping areas with high unemployment

Activities

1 a) Which two Channel ports does the tunnel run between?
  b) How long is the Channel Tunnel route?
  c) How are people and vehicles transported through the tunnel?

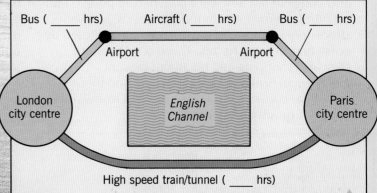

Bus (____ hrs)   Aircraft (____ hrs)   Bus (____ hrs)

Airport   Airport

London city centre   English Channel   Paris city centre

High speed train/tunnel (____ hrs)

E

2 a) Make a copy of diagram **E**.
  b) Using bar chart **C**, write in the times spent in the bus, aeroplane and train.
  c) With help from your completed diagram, say which would be the easiest and most relaxing journey to make between London and Paris. Give reasons for your answer.

3 Give three ways in which the tunnel project could:
  ● damage the environment
  ● help create jobs.

4 Suggest why the following people might be against the Channel Tunnel project.
  ● Villagers near Folkestone.
  ● Channel ferry crews.
  ● Farmers near the high speed train (TGV) route to Paris.

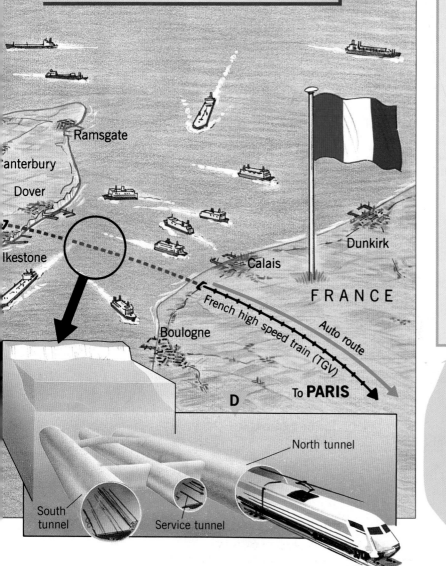

Ramsgate

Canterbury

Dover

Ilkestone

Boulogne

Calais

Dunkirk

F R A N C E

French high speed train (TGV)

Auto route

To **PARIS**

D

North tunnel

South tunnel

Service tunnel

How can a city street be improved?

Most main streets in UK towns and cities are busy and congested. They were built before people had cars and are unsuited to today's needs. Many streets are narrow and dangerous, and polluted by noise and exhaust fumes. There is no room to park and shopping can be an unpleasant experience. Buildings are often ugly and there is a lack of landscaping and open space. Overall, there is a poor **quality of environment**.

Drawing **A** shows a typical main street in a UK town with many of the problems just described. The people in the town want their street improved, and they have approached the local authority with their concerns. The authority agreed to produce a scheme that would:
1 reduce traffic congestion and pollution
2 make the area safer and more attractive
3 improve local shopping facilities.

Points are given for each feature.
For example:

- If a place is very attractive it will score 5 points.

- If it is ugly it will score 1 point.

- If it is in between it will score 2, 3 or 4 points.

The higher the number of points, the better the quality of environment.

QUALITY OF ENVIRONMENT SURVEY SHEET						
High quality 5 4 3 2 1 Low quality						
Attractive						Ugly
Quiet						Noisy
Tidy						Untidy
Safe						Dangerous
Few cars						Many cars
Easy movement						Congested
Good shopping						Poor shopping
Good parking						Poor parking
Open space						No open space
Like						Dislike

Place .... Total .... out of 50

Drawing **B** shows the proposed improvement scheme. This must be discussed, and alterations suggested, before building can begin.

1 Make a list of the problems shown in drawing **A**.

2 Now look carefully at the improvement scheme in drawing **B**.
   a) How has traffic congestion been reduced?
   b) What has been done to improve safety?
   c) What has been done to make the area more attractive?
   d) How has shopping been improved?

3 a) Make two copies of the survey sheet above.
   b) Complete a survey for drawing **A.**
   c) Tick the points you would give for each feature and add up the total number of points.
   d) Complete a similar survey for drawing **B**.

4 Use the two surveys to measure the success of the improvement scheme. What features still need to be improved? Suggest what could be done to make these better.

5 The views of local people must be considered. For each person in drawing **C**, say if they would be **for** or **against** the scheme. Give reasons for their views.

6 Should the scheme go ahead? Write a letter to the local authority giving your views and suggestions.

A The area before improvements

B Proposed improvement scheme

C

Mrs Briggs
A mother with two young children living on Main Street.

Mr Banks
Owner of a shop on Main Street. Often has to drive to his other shops in nearby towns.

Mr and Mrs Bell
Owners of a shop on Main Street which will be knocked down for parking.

6 The United Kingdom

Where is the UK?

The world can be divided into seven large landmasses and several large sea areas. The land masses, which cover 30 per cent of the Earth's surface, are called **continents**. The five largest sea areas are known as **oceans**. The continents and oceans are shown on map **A**.

The **United Kingdom**, more commonly referred to as the **UK**, is located to the north-west of the continent of Europe. It has joined together with 14 other European countries to form a group known as the **European Union** – or the **EU** for short. These countries, as shown on map **B**, lie mainly in western Europe. Many countries in eastern Europe are now trying to join the EU.

A The UK in the world

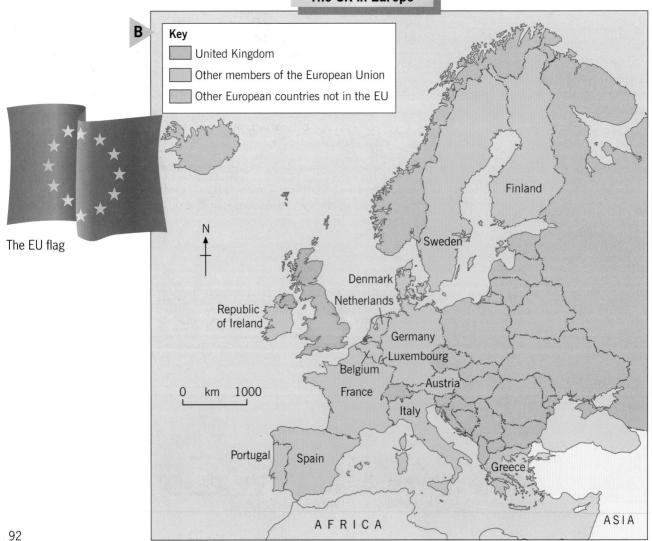

B The UK in Europe

The EU flag

What is the UK?

The United Kingdom (map **E**) is part of the **British Isles**. It consists of four countries:

- the three countries of England, Wales and Scotland that form **Great Britain** (map **D**), and

- the country of Northern Ireland.

Each country within the UK has a certain amount of self-rule and is able to develop its own distinct characteristics. These characteristics may include language and culture together with their own legal and education system.

You may have noticed earlier in this book how geographers like to group things together. Things may be grouped together to make them easier to describe or to understand. For example, the groups may be:

- types of settlement – dispersed, nucleated and linear (pages 56 and 57)

- land use in a city – the CBD, the inner city and the suburbs (pages 62 and 63)

- types of transport – road, rail, water and air (pages 76 and 77)

- causes of flooding (pages 38 and 39) or

- groups of countries – in the world, in the EU and in the UK (these pages 92 and 93).

C

The British Isles consist of two large islands. These islands are called Great Britain and Ireland.

D

Great Britain, the largest island, consists of three countries – England, Wales and Scotland. Ireland is divided into two countries – Northern Ireland and the Republic of Ireland.

E

The United Kingdom consists of the four countries of England, Wales, Scotland and Northern Ireland. The Republic of Ireland is an independent country.

Activities

1 What is the difference between **a)** the British Isles **b)** the United Kingdom and **c)** Great Britain?

2 Table **F** shows the area and population of five countries. Regroup the data under these headings:
a) British Isles **b)** UK **c)** England.

3 Group the following places under the headings: **a)** continents **b)** EU and **c)** UK.

Africa, Antarctica, Asia, Belgium, Denmark, England, Europe, Finland, France, Germany, Greece, Italy, Luxembourg, Netherlands, North America, Northern Ireland, Oceania, Portugal, Republic of Ireland, Scotland, South America, Spain, Sweden, UK, Wales.

F

	Area (km^2)	Population
England	130,423	49,089,000
Northern Ireland	14,160	1,663,300
Republic of Ireland	70,280	3,625,000
Scotland	78,133	5,128,000
Wales	20,766	2,921,000

Summary

The UK, which consists of England, Wales, Scotland and Northern Ireland, is a country in Europe.

What are the UK's main physical features?

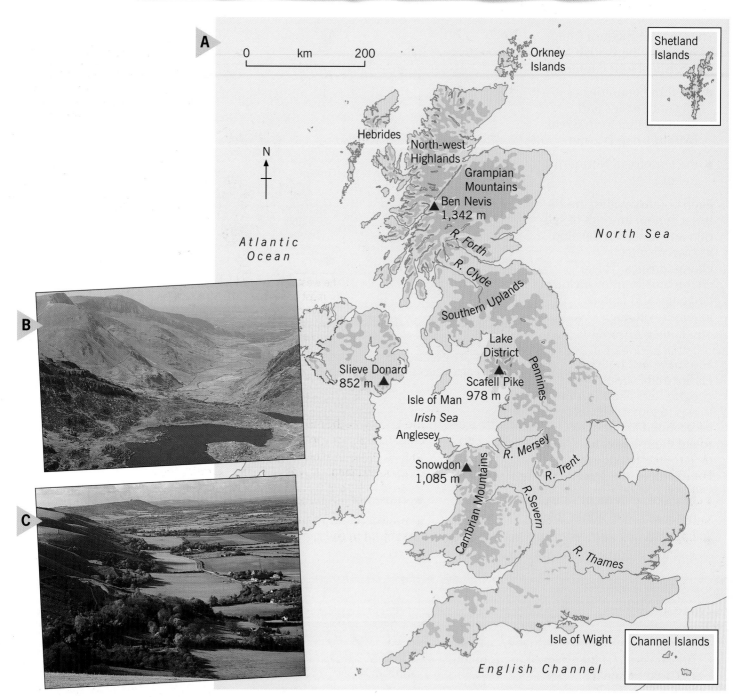

Map **A** above is a **physical map** of the UK. It shows the main sea areas, the largest islands, the longest rivers, the largest mountain areas and the highest mountains.

Most of the north and west of the UK is mountainous (photo **B**). This is because the rocks found here are old and tend to be resistant to erosion. The area has, over millions of years, been slowly worn away to leave rugged mountain peaks and deep, often lake-filled, valleys.

In contrast, much of the south and east of the UK is low-lying (photo **C**). The typical scenery is one of wide and flat-floored river valleys separated by areas of rolling hills.

Nowhere in the UK is far from the sea. The coastline, which is over 11,000 km in length, has scenery which varies from high cliffs to low-lying sand dunes.

For such a small country, there are also considerable differences in weather and climate between different places in the UK. Temperatures are usually higher in the south of England than they are in the north of Scotland (map **D**). Rainfall, which can fall at any time of the year, is often greater in the west of Britain than it is in the east (map **E**).

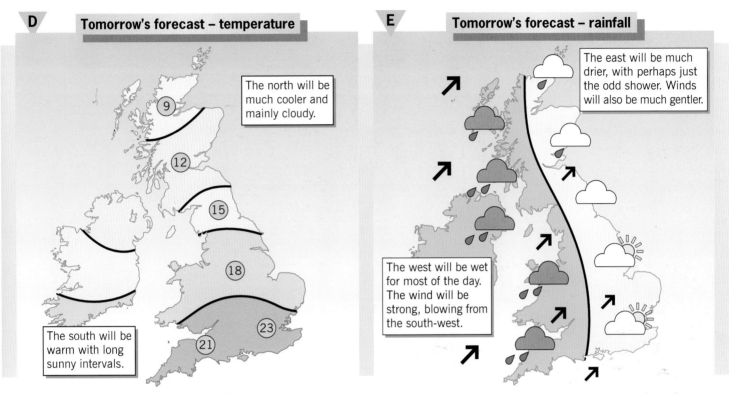

D Tomorrow's forecast – temperature

The north will be much cooler and mainly cloudy.

The south will be warm with long sunny intervals.

E Tomorrow's forecast – rainfall

The east will be much drier, with perhaps just the odd shower. Winds will also be much gentler.

The west will be wet for most of the day. The wind will be strong, blowing from the south-west.

The UK is said to have a **variable** climate because the weather often changes from day to day (if you have made daily recordings of the weather you will have noticed this for yourself). The climate is also said to be **temperate**, with warm summers, mild winters and some rain falling throughout the year. It is unusual for the UK's weather to be either too hot, too cold, too dry or too wet. However, although the UK usually avoids such weather hazards as drought, heat waves or extreme cold, the frequency and severity of flooding (Chapter 3) and storms does seem to have increased in recent years. The UK's weather is described in more detail in Chapter 2.

Activities

1 **a)** Look at photo **B** and the maps **D** and **E**.
   - Where in the UK was the photo taken?
   - Describe the scenery in the photo.
   - Describe the climate of the area.
   - Would you like to live there? Give reasons for your answer.
   b) Look at photo **C** and maps **D** and **E**. Answer the same questions as you did in part **a**.

2 Maps **D** and **E** show opposites: warm and cool, wet and dry.
   a) Using these two maps, which parts of the UK are:
   - wettest and driest throughout the year
   - warmest and coolest in summer
   - mildest and coldest in winter
   - windiest and least windy?
   b) Can you see any links between the areas of highest land (mountains) and lowest land, as shown on map **A**, and the answers you gave for part **a**?

Summary

Both the scenery and the climate of the UK are varied. Usually the scenery is very attractive while the climate is temperate and without extremes.

What are the UK's main human features?

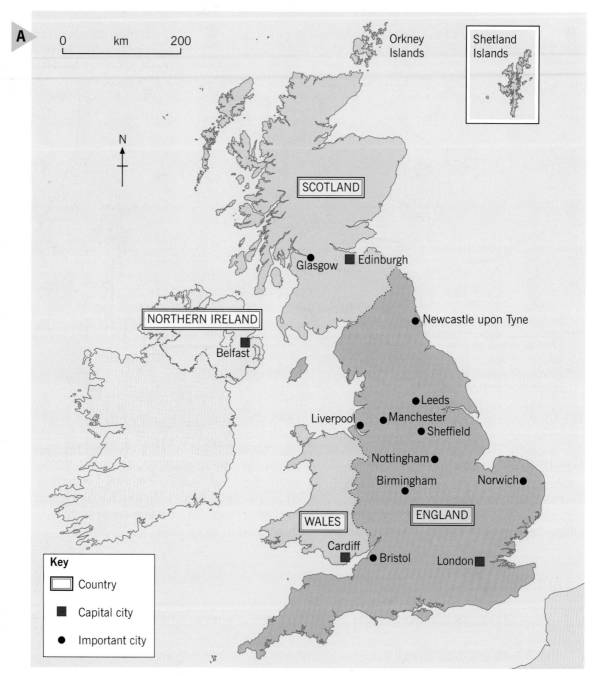

A

The map above is a **political map** of the UK. It shows the four countries, the four capital cities and some of the more important cities within the UK.

If you are living permanently in the UK, you will notice that your passport gives your nationality as British. However, that does not mean that everyone who is British is the same. There are often distinct differences between people living in each of the four UK countries. This is because different groups of people tend to develop their own customs and way of life. This can include their language and religion, how they dress and behave, what they eat, and what they do in their spare time. This creates a sense of **identity**.

The Scots, the Welsh and the Irish have all developed their own sense of identity. If we put their characteristics together we get a mental picture of 'typical' Scots, Welsh or Irish people. We call this image a **stereotype**. We must remember, though, that there are many types of Scots, Welsh and Irish people – they are not all the same!

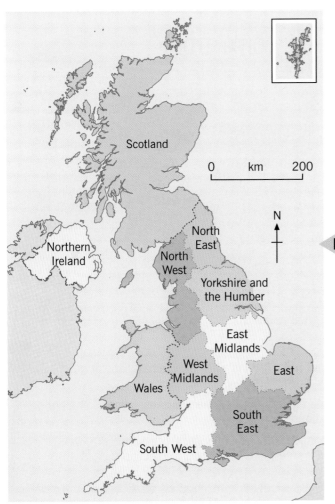

Map **B** shows the major **regions** of England and Wales. There are two types of region:

1 Places with similar physical characteristics, for example:
- the Lake District and the Cambrian Mountains are highland regions
- western Britain is a region with heavy rainfall.

2 Areas that have been created for administration purposes. In the UK this includes economic regions (map **B**), counties (Activity 2), and parliamentary constituencies. Most administrative regions have boundaries that are not natural but have been created by people.

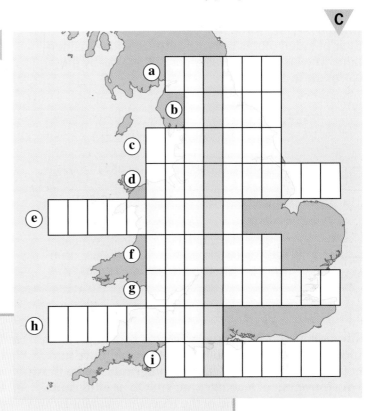

Activities

1 Ten **English** cities are shown on map **A**. Complete quizword **C**, using the clues below to name nine of these cities. The letters in the shaded boxes spell out the missing city. You may have to refer to map **A** on page 94, or an atlas, to answer some clues.
   a) A city on the River Thames
   b) A large city east of the Pennines
   c) The largest city in East Anglia
   d) A large city west of the Pennines
   e) The largest city in the West Midlands
   f) A city in the South West region
   g) A city on the River Trent
   h) A city at the mouth of the River Mersey
   i) A city in the Yorkshire and Humberside region

2 a) What is the name of the city in the shaded boxes of the quizword?
   b) Make up two clues for this city.

3 Make a list of the nine economic regions in England as shown on map **B**. For each region, choose three administrative areas from the 46 listed in box **D**.

D ▷

Avon, Bedfordshire, Berkshire, Buckinghamshire, Cambridgeshire, Cheshire, Cleveland, Cornwall, Cumbria, Derbyshire, Devon, Dorset, Durham, East Sussex, Essex, Gloucestershire, Greater London, Greater Manchester, Hampshire, Hereford and Worcester, Hertfordshire, Humberside, Isle of Wight, Kent, Lancashire, Leicestershire, Lincolnshire, Merseyside, Norfolk, Northamptonshire, Northumberland, North Yorkshire, Nottinghamshire, Oxfordshire, Shropshire, Somerset, South Yorkshire, Staffordshire, Suffolk, Surrey, Tyne and Wear, Warwickshire, West Midlands, West Sussex, West Yorkshire, Wiltshire.

Summary

Most people in the UK live in England. The UK can be subdivided into countries, economic regions and administrative regions.

Where do people in the UK come from?

The movement of people from one place to another is called **migration**, and a person who moves is known as a **migrant**. People, whether in groups or as individuals, have always moved. Table **A** explains how some people may have moved because they wanted to, and others because they had no choice and just had to move.

Many of the early migrants into Britain came as **invaders** (map **B**). While some may have returned home with their plunder, others became **settlers** and remained here. The last major invasion of Britain was by the Normans in 1066.

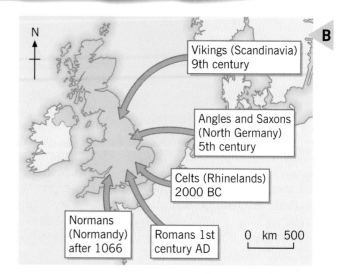

A

Voluntary migration is when people choose to move. This may be:	**Forced migration** is when people have little or no choice but to move. This may be the result of:
<ul><li>to improve their standard of living, e.g. to find more or better-paid jobs</li><li>to improve their quality of life, e.g. retiring to a better climate, living and working either in a more pleasant environment or where there are better services.</li></ul>	<ul><li>natural disasters, e.g. earthquakes, volcanic eruptions or floods</li><li>economic, social and political pressures, e.g. war, famine, religious or political persecution.</li></ul>

The British population in the 20th century

Throughout the 20th century, people continued to move into and out of the UK. People who move into a country like the UK are called **immigrants**, while those who move out of a country are known as **emigrants**. Diagram **C** shows some of the recent immigrant groups into the UK, and gives reasons why they came to settle here.

The effect of migration into the UK has been considerable. It has caused an increase in numbers and changed the mix of people in the country (graph **D**). It has produced a **multicultural society** where people of different ethnic groups, language, religion and culture live and work together (photo **E**). While this mix can cause problems, most people agree that it has added variety and created benefits.

C

A large number of Chinese also now live in the UK.

My grandparents came from the West Indies in the 1950s. They were encouraged to come by the British government as the UK had a serious labour shortage after the Second World War.

Many generations of Irish have come to Britain, mainly to find jobs.

Many immigrants came from Eastern Europe in the 1940s and 1950s to escape war and religious and political persecution.

Many immigrants to Britain have come from India, Pakistan and Bangladesh. They came to Britain to seek work and a better education, to start their own businesses and to improve their standard of living.

D Ethnic groups in the UK, 1997

Black Afro-Caribbean 1.7%   Chinese and other Asian 0.7%
Indian, Pakistani and   Others 0.7%
Bangladeshi 2.9%

White 94%

E

Recent immigrants into the UK

Diagram **F** shows that, during the 1990s, three out of every five immigrants into the UK came from the EU. This was mainly due to the ease of moving from one EU country to another. By 2000, however, certain people in the UK were becoming increasingly worried about three types of immigrant:

- **Refugees** who come to Britain because they might be tortured, imprisoned or killed in their home country. This persecution may be due to their ethnic grouping or their religious or political beliefs, e.g. people from the former Yugoslavia.

- **Asylum seekers** who claim to be refugees but who have to prove their claim, e.g. the hijackers and passengers of a plane from Afghanistan.

- **Illegal immigrants** who try to enter the UK illegally, e.g. the 56 Chinese who suffocated in a lorry as it came across the Channel.

F Source of the most recent migrants into the UK

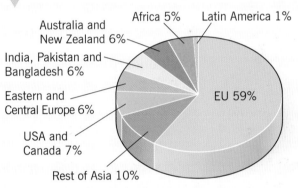

Africa 5%   Latin America 1%
Australia and New Zealand 6%
India, Pakistan and Bangladesh 6%
Eastern and Central Europe 6%
USA and Canada 7%
Rest of Asia 10%
EU 59%

Activities

1 **a)** Make a list of where everyone in your class was born. Plot the information on a map of the UK or, if many were born outside the UK, on a map of the world.
   b) Draw two types of graph to show where people in your class were born. Pages 14 and 15 will help you decide which types of graph to use. You may be able to draw your graphs on a computer. The graphs might show those born in each of the following places:
   - within 10 km of your school
   - over 10 km from the school but within the same region or county
   - elsewhere in the UK
   - in one of the other 14 EU countries
   - elsewhere in the world.

2 Your class should divide into four or five groups. Each group should then investigate one of the groups of people who have moved into the UK since 1900, e.g. West Indians, or refugees from the former Yugoslavia. The questions on clipboard **G** should help you with your investigation.

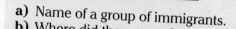

G
a) Name of a group of immigrants.
b) Where did they come from?
c) Why did they leave their home country?
d) When did they come to the UK?
e) Why did they come to the UK?
f) Once in the UK, where did most of them settle?
g) Why did they settle there?
h) What contribution have they made to the community in which they live?

Summary

The UK is made up of people from many different countries. Different ethnic groups and differences in language, religion and culture have given the UK a multicultural society.

What jobs do people in the UK do?

As you know, there are hundreds of different types of jobs. Geographers often call jobs **economic activities** because people earn money from them. It is convenient, when looking at economic activities, to group the jobs that people can do into three main types. These are explained in figures **A**, **B** and **C.**

A

Primary activities are those where people collect and use natural resources. These resources may be obtained from the land or the sea. Examples of primary activities are:

● farming

● fishing

● forestry ⟶

● mining and quarrying.

B

Secondary activities are those where people:

● make, or manufacture, things from natural resources, e.g. producing steel from iron ore, or making marmalade from oranges

● assemble or construct things that have already been made from natural resources, e.g. assembling cars, or building new houses. ⟶

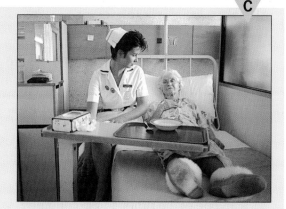
C

Service activities, sometimes called **tertiary** activities, are those that help people to improve their quality of life. These activities include:

● education

● health ⟶

● retailing (shopping)

● entertainment

● transport

● tourism.

As diagram **F** shows, the main types of economic activity in the UK have changed over a long period of time:

● Before 1800 (photo **D**), most British people were employed in primary activities.

● During the 1800s and early 1900s (photo **E**) the number employed in primary activities began to decline, while those in secondary activities increased rapidly. This followed the invention of the steam engine.

● From the late 1900s to the present day, the number in both primary and secondary activities decreased, while those providing a service increased.

D Gainsborough's painting (1750) shows Mr and Mrs Andrews and a farming scene

E Lowry's painting shows factories in Lancashire in the 1930s

F

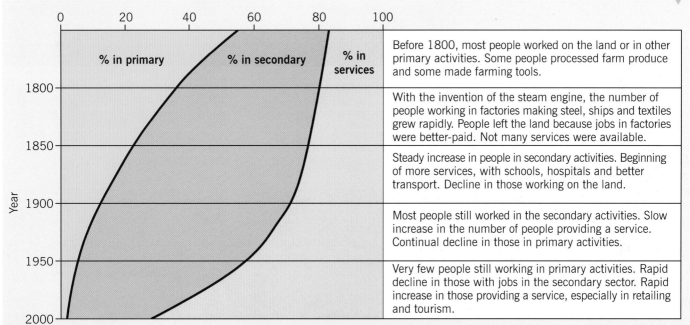

% in primary	**% in secondary**	**% in services**	Before 1800, most people worked on the land or in other primary activities. Some people processed farm produce and some made farming tools.
			With the invention of the steam engine, the number of people working in factories making steel, ships and textiles grew rapidly. People left the land because jobs in factories were better-paid. Not many services were available.
			Steady increase in people in secondary activities. Beginning of more services, with schools, hospitals and better transport. Decline in those working on the land.
			Most people still worked in the secondary activities. Slow increase in the number of people providing a service. Continual decline in those in primary activities.
			Very few people still working in primary activities. Rapid decline in those with jobs in the secondary sector. Rapid increase in those providing a service, especially in retailing and tourism.

Activities

1 **a)** What is meant by the term 'economic activities'?
   b) What are primary, secondary and service activities?
   c) Give three examples of each of primary, secondary and service activities.
   d) What are the main types of job in your local area? Sort them into primary, secondary and services.

2 Look at diagram **F**.
   a) In which year were there most primary jobs?
   b) In which year were there most secondary jobs?
   c) In which year were there the least primary jobs?
   d) What percentage of the total jobs were service activities in 2000? Suggest a reason for this.

3 Would you be likely to see today the landscapes shown in photos **D** and **E**? Suggest reasons for this.

Summary

Jobs, or economic activities, may be divided into three main groups: primary, secondary and services. The proportion of people working in each group changes over time.

101

What are the differences between urban and rural areas?

Urban and **rural** are two words that are frequently used by the media. The simplest meaning of each is that urban is to do with the city and rural is to do with the countryside. When you hear each term, what picture comes to your mind? Is it like the painting by Canaletto (1740s) in picture **A** or by Constable (1810s) in picture **B**?

Your picture is likely to be your **perception** of an urban or a rural area. By perception, we mean what we imagine a place to look like. Sometimes our perception of a place can be quite accurate, especially if we live there or have visited it. But often our perception is very different from reality. Many people who live in urban areas have a wrong perception of what living in a rural area is really like, just as many people living in rural areas have a wrong perception of what it is like to live in an urban area.

Photo **C** shows part of a city centre. Photo **D** is an area of countryside. Do these photos, together with the comments made in figures **E** and **F**, confirm or change your perception of urban and rural areas?

A

B

C

D

E

We like living in a city because:
1 Lots of jobs with higher pay.
2 Near to schools, hospitals and shops.
3 Lots of entertainment such as cinemas, discos and pubs.
4 Good transport between places.

We would not like to live in a village because:
1 Too quiet – nothing to do.
2 Away from our friends – too isolated.
3 No public transport to get into town. Need your own car.
4 Smells of manure, and farm animals make roads dirty.

We live in a big city.

F

We like living in a village because:
1 Plenty of open space – countryside and gardens.
2 Not much traffic.
3 Cleaner air and more attractive scenery.
4 Better community spirit.

We would not like to live in a big city because:
1 Too many people and too much traffic – congested.
2 Dirty, e.g. litter.
3 Smells of car fumes.
4 Very noisy.

We live in a small village.

Activities

1 a) What do we mean by the term 'perception'?
 b) What was Dick Whittington's perception of London?

2 a) Give the meanings of **urban** and **rural**.
 b) Make two copies of figure **G**. Label the first copy 'urban' and the second 'rural'.
 c)
> On each sheet put a tick ✓ in each horizontal row:
> ✓ 5 if the place is **attractive**
> ✓ 4 if the place is **fairly attractive**
> ✓ 3 if the place is **neither attractive nor ugly**
> ✓ 2 if the place is **quite ugly**
> ✓ 1 if the place is **ugly**
> The higher the number, the better the quality.

 d) Add up the total ticks for each column. Which has the higher quality of the environment?
 e) In which of the two environments would you prefer to live? Give three reasons why you would prefer to live there and three reasons why you would not want to live in the place that you rejected.

G

QUALITY OF THE ENVIRONMENT ASSESSMENT SHEET

← High quality   Low quality →
5 4 3 2 1

Attractive						Ugly
Interesting						Dull/boring
Quiet/peaceful						Noisy/busy
Clean/tidy						Dirty/untidy
Safe						Dangerous
No cars						Many cars
Open space/play areas						No open space
Good shopping						Poor shopping

Place .... Total .... out of 40

Summary

The perception by urban dwellers of rural areas, and by rural dwellers of urban areas, is often inaccurate. These inaccuracies can cause problems.

What are the regional differences in the UK?

A

The map on this page is a **satellite image** of the UK. A satellite image is a photo that is taken from space and sent back to Earth. Many satellite images have 'false colours'. This means that colours may be changed slightly to highlight, or enhance, certain characteristics, e.g. the extent of flooding, a weather system (page 24), or the effect of deforestation.

On this satellite image:

● dense forest and areas of lush vegetation show up as bright green

● highland and other places that lack vegetation appear as various shades of brown

● rivers and lakes are blue, and

● urban areas are pink.

We have already seen on page 97 that the UK can be divided into several regions. In map **B**, the number of regions has, for simplicity, been reduced to six. The main characteristics of relief, climate, settlement and economic activities for each of these regions, appear on map **A**. You will notice that each of these regions has:

● some similarities with other regions

● some differences from other regions.

Wales
● Mainly highland with mountains in the north.
● Heavy rainfall all year, especially in winter.
● Warm summers and mild winters in low-lying areas.
● Colder, with more wind, in highland areas.
● Most people live in the south, elsewhere is mainly rural.
● Secondary and service activities in the south, mainly primary elsewhere.

South and west of England
● Many hills and moorland areas.
● Heavy rain all year, especially in winter.
● Warm summers and very mild winters.
● Mainly rural with a few urban centres.
● Mainly primary and service activities.

Scotland
- Mainly mountainous.
- Central area, with the rivers Forth and Clyde, is lower.
- Heavy rainfall all year, especially in the west.
- Highest areas get snow and strong winds in winter.
- Cool summers, mild winters.
- Most people live in central areas, elsewhere is mainly rural.
- Secondary and service activities in central areas, mainly primary with some service activities in other areas.

North and east of England
- Low-lying apart from the Pennines in the west.
- Light rainfall all year.
- Cool summers, cold winters.
- Many large urban areas.
- Decline in primary and secondary activities, increase in service activities.

North and west of England
- Includes the highlands of the Lake District and Pennines.
- Heavy rainfall all year, especially in highland areas and in winter.
- Cool summers, mild winters.
- Mainly urban to the south and west of the region, mainly rural to the north and east.
- Secondary activities in the south, primary activities in the north and east, services more widespread.

South and east of England
- Mainly low-lying with gentle hills.
- Several long rivers, such as the Thames.
- Some rain all year, most in summer.
- Cold winters, warm summers.
- Both urban and rural.
- Many service activities, few primary or secondary jobs.

Activities

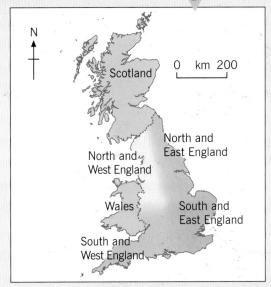

B

1 Match the following regions with the best descriptions:

Regions	Descriptions
Scotland	is mainly flat and low-lying.
Wales	is mountainous with a central lowland.
North and west England	is the mildest region.
North and east England	includes the highest land in England.
South and east England	is hilly and has cold winters.
South and west England	has mountains in the north.

2 a) Make a FactFile to show the main characteristics of your local area or region. Use the headings shown in diagram **C**.

   b) Make another FactFile using the headings in FactFile **C**. This time, choose a different region to the one in which you live. Say in what ways this region is similar or is different from yours.

C

FactFile
- Relief ...
- Climate ...
- Settlement ...
- Economic activities ...

Summary

The UK can be divided into several regions. Each region has its own characteristics of climate, relief, settlement and economic activities.

The UK enquiry

This enquiry is concerned with your ability to use a range of resources, your understanding of what England and Wales is like, and what you have learned from this unit on the UK.

Your task is to plan a seven-day sightseeing tour of England (and Wales if you wish) for a pen-pal who lives in either the EU or the USA and is coming to the UK to visit you. Your tour should include visits to places that are both interesting and typical of your country. Your pen-pal is due to:

● arrive at either your local airport or one of London's airports on a Monday morning

● depart from London airport on the following Sunday afternoon.

What are some of England's most interesting places?

Your task should be divided into three main parts.

1 You will need an introduction in which you explain what the enquiry is asking you to do.

2 You will have to collect various resources and explain how they will help you plan the tour.

3 You will need to describe the tour. This should be done in diary form. The diary should include reasons for your choice of places to be visited, and a reference as to how you will travel from one place to the next. This section should be illustrated with labelled maps and photos.

A

1 Introduction

You should decide:

● which places your pen-pal is most likely to want to visit

● which places will give your pen-pal a good idea of what England is really like.

You should try to include visits to:

● a rural area that has attractive scenery – this could be mountains, lakes, a river valley or a stretch of coastline

● an urban area where there are lots of interesting places to visit

● a historic building, e.g. a castle or a cathedral or a historic site (for example a scene of a battle or famous event)

● a place with a lots of entertainment.

2 Planning the tour

This will allow you to use a wide range of resources. Useful resources might include:

- national bus and rail timetables
- rail, bus and car route maps
- CD-ROMs such as Encarta or AutoRoute Express
- the internet
- holiday brochures.

Remember: You must be practical in terms of time and money. You only have seven days and so cannot visit everywhere in the country. You also need to be careful with money – you cannot fly everywhere!

B Differences between places on map C

BIRMINGHAM									
181	CAMBRIDGE								
172	330	CARDIFF							
317	413	480	KESWICK						
192	96	248	501	LONDON					
142	258	309	192	128	MANCHESTER				
333	368	510	142	458	230	NEWCASTLE			
109	160	174	434	91	258	413	OXFORD		
325	469	261	632	386	461	662	312	PLYMOUTH	
214	251	392	184	339	114	139	296	544	YORK

London to Oxford = 91 km

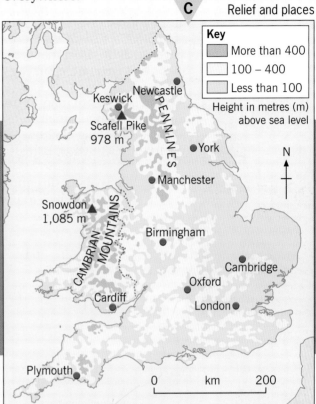

C Relief and places

Key
- More than 400
- 100 – 400
- Less than 100

Height in metres (m) above sea level

N

Newcastle, Keswick, Scafell Pike 978 m, PENNINES, York, Manchester, Snowdon 1,085 m, CAMBRIAN MOUNTAINS, Birmingham, Cambridge, Oxford, Cardiff, London, Plymouth

0   km   200

3 Writing up the tour

- Write up, in diary form, your proposed five-day tour. You could set this out as in table **E** below.
- Draw a map to show your proposed route. Use different colours for different types of transport.
- Justify your choice of places that you plan to visit.

E

	Places to be visited	Reasons for visit	Travel between places	Distance between places	Overnight stay
Day 1 Monday	Arrive at ...				
Day 2 Tuesday					
Day 3 Wednesday					
Day 4 Thursday					
Day 5 Friday					
Day 6 Saturday					
Day 7 Sunday	Fly from London airport ...				

Follow-up work

- Compare your tour plans with those of others in your class. In what way are the tours:
  - **a)** similar
  - **b)** different?
- As a class you could then:
  - **a)** decide who produced the most varied and interesting route, and then
  - **b)** produce a combined effort, perhaps for a wall display.

How can we show direction?

Maps show what things look like from above. They are very useful because they give information and show where places are. There are many different types of map. These include street maps, road maps, **atlas** maps and **Ordnance Survey** (OS) maps.

A **plan** is a type of map. Plans give detailed information about small areas. Places like schools, shopping centres, parks and leisure centres are shown on plans.

This section is about **direction**. The best way to show direction is to use the **points** of the compass. There are four main points. These are north, east, south and west. You can remember their order by saying '**N**ever **E**at **S**hredded **W**heat'.

Between these four main points there are four other points. These are north-east, south-east, south-west and north-west.

Most maps have a sign to show the **north** direction. If there is no sign the top edge of the map should be **north**.

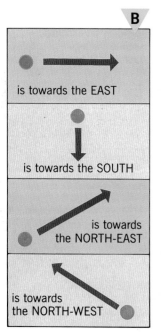

A

North

West — East

South

Four point compass

Never

Wheat — Eat

Shredded

Remember...

North

North-west — North-east

West — East

South-west — South-east

South

Eight point compass

To give direction for a place you have to say which way you need to go to get there. The direction is the point of the compass *towards* which way you have to go. Diagrams **B**, **C** and **D** show you how to give a direction.

B

is towards the EAST

is towards the SOUTH

is towards the NORTH-EAST

is towards the NORTH-WEST

C

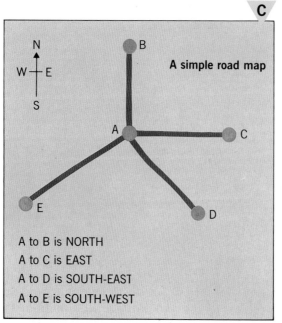

A simple road map

A to B is NORTH
A to C is EAST
A to D is SOUTH-EAST
A to E is SOUTH-WEST

D

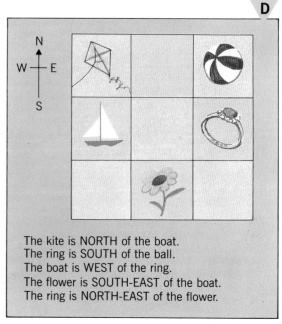

The kite is NORTH of the boat.
The ring is SOUTH of the ball.
The boat is WEST of the ring.
The flower is SOUTH-EAST of the boat.
The ring is NORTH-EAST of the flower.

Activities

1 Draw the compass in diagram **E** and label the unmarked points.

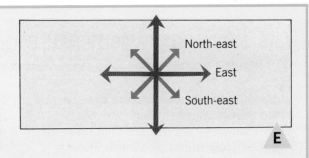

E

2 Copy these drawings and complete the sentences below them. The first one has been done for you.

B is north of **A** **D** is . . . of **C** **F** is . . . of **E** **H** is . . . of **G** **I** is . . . of **J**

F

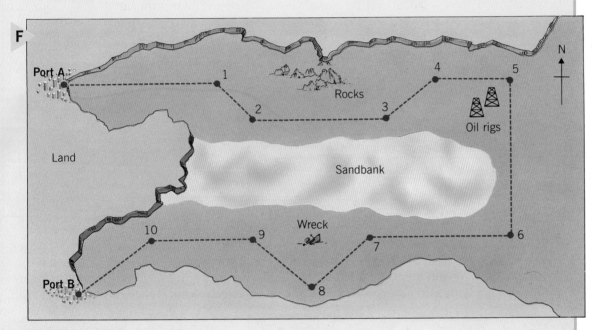

3 Study map **F** and give the following directions:
   a) from Port A to the rocks
   b) from the wreck to the oil rigs
   c) from the oil rigs to the rocks
   d) from the wreck to Port A
   e) from the rocks to the wreck.

4 a) A ship has landed its cargo at Port A. It must go to Port B to reload. The course the ship must follow is shown by the dotted line on the map. Give the Captain compass directions to follow between each numbered point. Start like this: *Leave Port A. Go east to point 1. Go south-east . . .*

   b) Imagine that the sandbank has been cleared to make ship movement easier. Work out the best course from Port B to Port A. Give compass directions to follow that course.

E X T R A

You will need to use the Ordnance Survey map of the Cambridge area for this question. It is on page 109.

Look at the villages near the bottom of the map. Give the following directions:
   a) from Foxton to Whittlesford
   b) from Foxton to Newton
   c) from Great Shelford to Whittlesford
   d) from Great Shelford to Haslingfield
   e) from Haslingfield to Harston.

Summary

Maps are a good way of giving information and showing where places are. Direction can be described by using the points of the compass.

How can we measure distance?

A map can be used to find out how far one place is from another. Maps have to be drawn smaller than real life to fit on a piece of paper. How much smaller they are is shown by the **scale**. This shows you the **real** distance between places. In diagram **A** the scale line shows that 1 cm on the map is the same as 1 km on the ground. Every map should have a **scale line**.

Straight line distances are easy to work out. Diagram **A** shows how to measure the straight line, or shortest, distance between the church and the bridge.

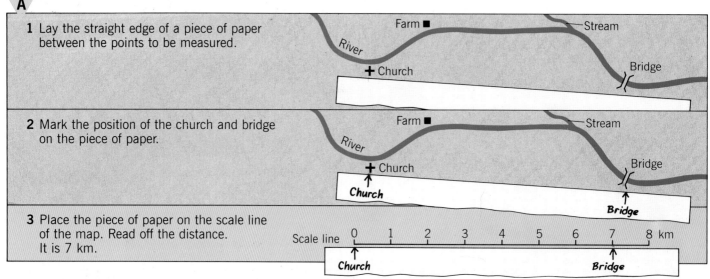

A

1 Lay the straight edge of a piece of paper between the points to be measured.

2 Mark the position of the church and bridge on the piece of paper.

3 Place the piece of paper on the scale line of the map. Read off the distance. It is 7 km.

The same method can be used to work out distances that are not straight lines. To measure these, divide the route into a number of sections and measure each one.

This can be done by using a piece of paper and turning it at each bend. Diagram **B** shows how to measure the distance from the church to the bridge, following the river.

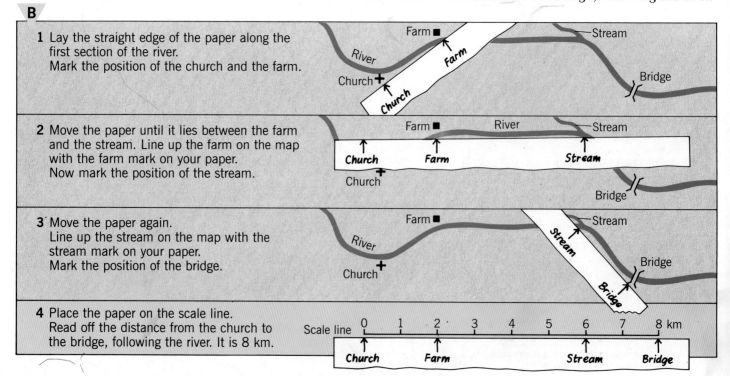

B

1 Lay the straight edge of the paper along the first section of the river.
Mark the position of the church and the farm.

2 Move the paper until it lies between the farm and the stream. Line up the farm on the map with the farm mark on your paper.
Now mark the position of the stream.

3 Move the paper again.
Line up the stream on the map with the stream mark on your paper.
Mark the position of the bridge.

4 Place the paper on the scale line.
Read off the distance from the church to the bridge, following the river. It is 8 km.

Activities

1 Use the scale line from map **C** to give the lengths of these lines. Answer like this:
Line **a)** is … metres (m) in length.
a) _____
b) _____
c) _____
d) _____

2 Use map **C** and the scale line to give the straight line distance between the places below. Choose your answers from the following:

| 40 m | 80 m | 100 m | 120 m |

a) Kate's house and the school.
b) Joanne's house and the post office.
c) Tim's house and the post office.
d) John's house and the garage.

3 a) Give the distance Joanne has to travel to school if she calls on Kate on the way.
b) Give the distance John has to travel to school if he calls at the shop and post office first.

4 What is the distance around the duck pond if you walk on the footpath? Give your answer in metres (m).

5 You have been given a map and instructions to help you find some hidden treasure. Follow the instructions to find out where it is.

Check the exact spot by sorting out the jumbled words in the treasure chest and choosing the correct answer.

Leave the wreck and go east for 2 km. Go north for 2 km, west for 1 km, then north-west for another 1 km. Now head north-east for 3 km and east for 2 km. Go south for 4 km and finally west for 1½ km.

NI EHT SLILH

REDUN ETH GIDBER LWOBE HET GIB ERTE

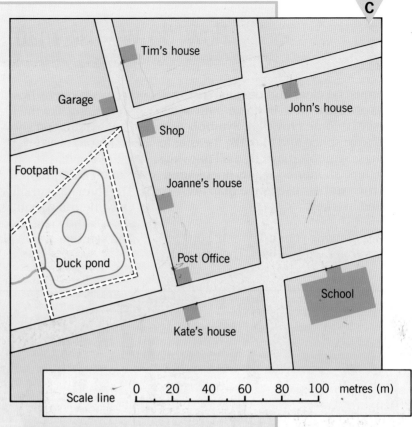

Summary Distances on a map can be measured using the scale line. The scale line gives the real distance between places on the map.

How do we use map symbols?

A map must be clear and easy to read. There is always a lot to put on a map and it can easily become crowded. **Symbols** are used to save space and to make it easier to see things. Symbols may be small drawings, lines, letters, shortened words or coloured areas. The symbols used on a map are explained in a **key**.

If you are drawing your own map, you can make up your own symbols. They should be as simple as possible and look something like the feature they stand for. How would you show a post box, a library or a football ground?

Sketch **A** and map **B** show the same street. The map has simplified the street scene. Only the main features of the street are shown and symbols are used to save space. The symbols are explained in the key.

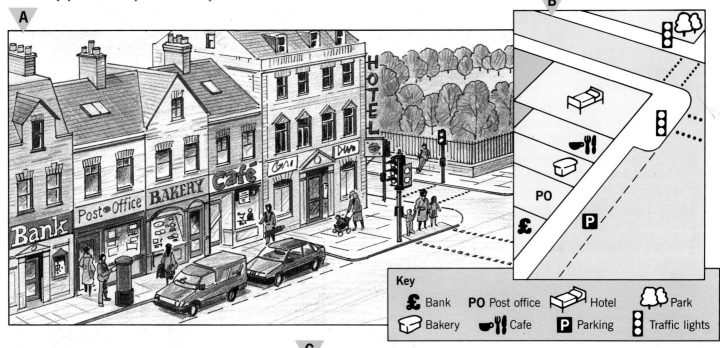

A

B

Key

£ Bank   PO Post office   🛏 Hotel   🌳 Park

🍞 Bakery   🍴 Cafe   P Parking   Traffic lights

The **Ordnance Survey** (OS) is responsible for mapping Britain. The OS produces very accurate maps that have a lot of information on them.

There is an Ordnance Survey map of the Cambridge area on page 125 of this book. The symbols used on that map are on page 124.

Look at the photos in **C**. They show some of the symbols used on Ordnance Survey maps. Which symbols could you work out without the answers being given?

C

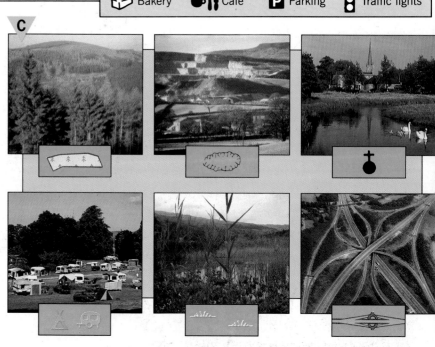

Activities

1 Look at map **D**. It is part of the Ordnance Survey map of the Cambridge area from page 125. It has been enlarged to make it easier to read. The scale has changed so the 4 cm on the map equals 1 km on the ground.

a) Make a copy of table **E** below.

b) Draw the symbols from map **D** in the correct columns of your table.

c) Say what each symbol shows. You will need to use the key on page 124. Some have been done for you.

© Crown Copyright

E

Drawings	Lines	Abbreviations (letters/shortened words)	Coloured aeas
\\\\//⁄ ⁄/⁄|\\\\ = Embankment	‿‿ = Contour	**Cemy** = Cemetery	▢ = Buildings

F

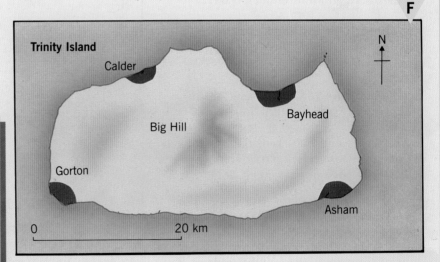

2 Make a larger copy of map **F**. It should be at least half a page in size. Using the Ordnance Survey symbols from page 124, draw on the map the following information.

◆ There is a main road between Gorton and Bayhead, and a second class road between Bayhead and Asham. A minor road joins Asham and Gorton and goes on to Calder.

◆ A railway line runs from Asham to Gorton, to Calder and on to Bayhead. The station at Calder is closed but the others are open.

◆ Gorton has a church with a spire and a chapel. Bayhead has a church with a tower and a post office. Asham has a telephone box and a youth hostel.

◆ The spot height at Big Hill is 312 metres high. The land south of Big Hill is marshy.

◆ The River Bee rises to the north of Big Hill and flows into the sea at Calder. (Remember to use bridges.)

◆ There is a wood on the east coast.

3 Draw a map of an island of your own. Use at least **15** different symbols. Name your towns, villages and other main features. Give your map a title.

Summary

Symbols are simple drawings that show things on maps. All maps have a key to explain the symbols.

113

What are grid references?

Maps can be quite complicated and it may be difficult to find things on them. To make places easier to find, a grid of squares may be drawn on the map. If the lines making up the grid are numbered, the exact position of a square can be given.

On Ordnance Survey maps these lines are shown in blue and each has its own special number. The blue lines form **grid squares**. **Grid references** are the numbers which give the position of a grid square.

On these two pages you will learn about **four figure grid references**.

To *give* a grid reference is simple. Look at the grid in diagram **A** and follow these instructions to give the reference for the yellow square.

- Give the number of the line on the *left* of the yellow square – it is 04.
- Give the number of the line at the *bottom* of the yellow square – it is 12.
- Put the numbers together and you have a four figure grid reference. It is 0412.

In the same way, the Picnic Square has a reference of 0313 and the Church Square is 0512.

What will be the grid references for the Bridge Square and the Tent Square?

To *find* a grid reference is also easy. Look at the grid in diagram **B** and follow these instructions to find grid square 4237.

- Go along the top of the grid until you come to 42. That line will be on the *left* of your grid square.
- Go up the side of the grid until you come to 37. That line will be at the *bottom* of your square.
- Now follow those two lines until they meet. Your square will be above and to the right of that point. There is a house in it.

What is in squares 4136 and 4037?

A

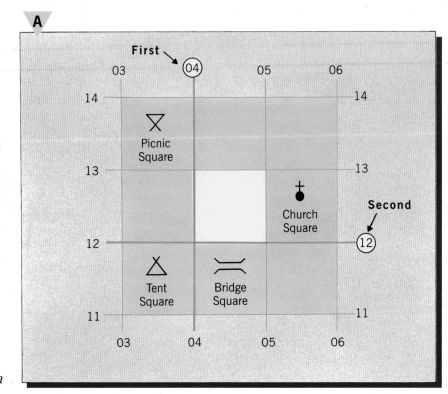

B

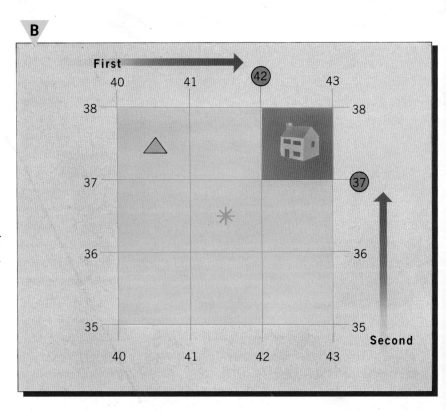

Activities

Look at map **C** of the British Isles. It shows some of the main towns, mountain areas and the three longest rivers. Use the map to answer the questions below.

✔ **Remember**

✔ The line on the left comes first.
✔ The line at the bottom comes second.

It may help you to remember if you say '**Along** the corridor and **up** the stairs'.

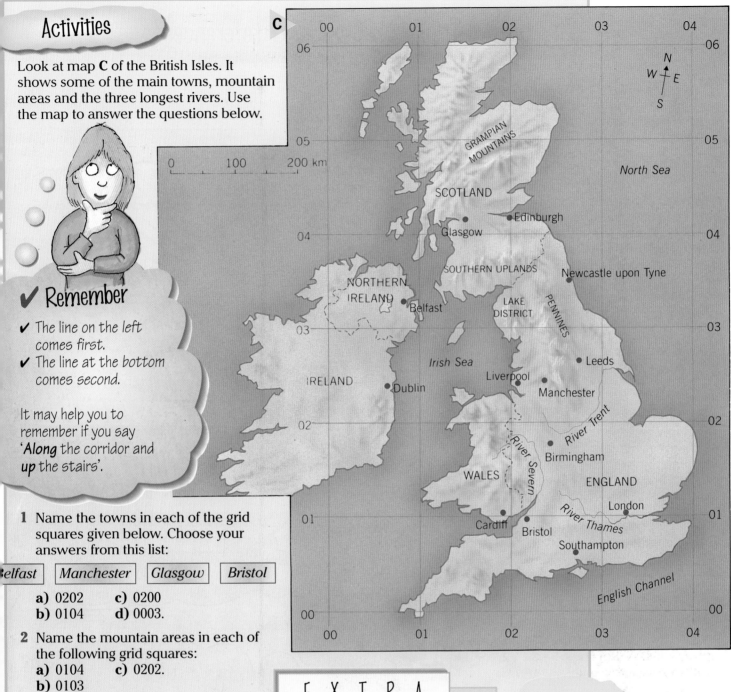

1 Name the towns in each of the grid squares given below. Choose your answers from this list:

| Belfast | Manchester | Glasgow | Bristol |

   a) 0202   **c)** 0200
   b) 0104   **d)** 0003.

2 Name the mountain areas in each of the following grid squares:
   a) 0104   **c)** 0202.
   b) 0103

3 **a)** Which rivers flow through grid square 0201?
   b) Which river reaches the sea in grid square 0201?

4 Give the grid references for these places:
   a) Dublin
   b) Newcastle upon Tyne
   c) London
   d) the Irish Sea.

5 Give the grid reference for the place where you live.

EXTRA

Look at the Ordnance Survey map on page 109. Name the farms in each of the following grid squares (the symbol for farm is Fm).
   a) 4149
   b) 4156
   c) 4456
   d) 4650
   e) 4257.

Summary

Grid references can be used to help describe the location of a place on a map.

How do we use six figure grid references?

Grid references are very useful in helping us to find places on maps. A four figure reference on an Ordnance Survey map equals an area on the ground of one square kilometre. This is quite a large area. To be more accurate we need to use a **six figure grid reference**. This pinpoints a place exactly to within 100 metres.

Look at the grid in diagram **A**. The six figure grid reference for the church is 045128. Follow these instructions and look at diagrams **B** and **C** to see how that reference is worked out.

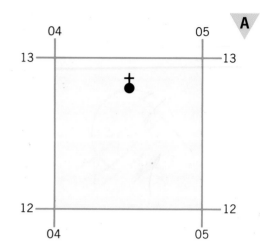

- Give the number of the line on the *left* of the yellow square – it is 04.

- In your head divide the square into tenths as shown in the grid in diagram **B**. Follow arrow **A** across the square. The church is about halfway across from the left. That puts it on the five-tenths line. Write down 5 after your number 04.

- You now have the first half of your six figure reference – it is 045.

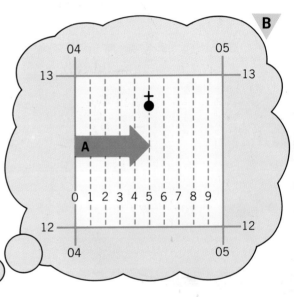

✔ Remember

✔ The numbers along the **bottom** come first.
✔ The numbers on the **left** come second.
✔ There must always be six figures.

- Now give the number of the line at the *bottom* of the yellow square – it is 12.

- In your head divide the square into tenths as shown in the grid in diagram **C**. Follow arrow **B**. The church is over halfway up from the bottom. That puts it on the eight-tenths line. Write down 8 after your number 12.

- You now have the second half of your six figure reference – it is 128.

- Put the two halves together and you have 045128.

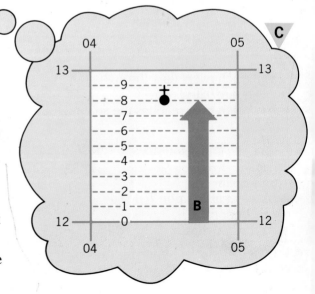

Activities

Look at map **D**. The 'tenths' lines have been added to help you with questions **1**, **2** and **3**. Check your references . . .

- The village of Eldon is in grid square 1623.
- The Mill is at reference 166256.
- Dingle Farm is at reference 170238.

1 Copy and complete the sentences below. Use the correct answer from the brackets.
   a) At 168245 there is a (church, post office, farm).
   b) At 165257 there is a (telephone, school, bridge).
   c) At 175233 there is a (farm, lake, level crossing).
   d) At 177244 there is a (station, wood, roundabout).

2 Give the six figure grid reference for each of the following:
   a) Eldon post office
   b) Causey railway station
   c) Padley school
   d) Burr Wood picnic site.

3 a) Follow these directions for a pleasant walk:

   Start at 170238. Walk down to 173237. Turn left and go to 177244. Go along the road to 171248. Follow the path to 178257. Turn left and finish your walk when the path reaches the road.

   b) Name the place where you finished your walk. Give its six figure grid reference.
   c) Where would you have stopped for lunch?
   d) How many churches did you pass on the way? Give their six figure grid references.

4 You will need to use the Ordnance Survey map of the Cambridge area for this question. It is on page 125.
   a) Make a copy of table **E**.
   b) Use the map to complete table **E**. The missing symbols, meanings and references are given in diagram **F**.

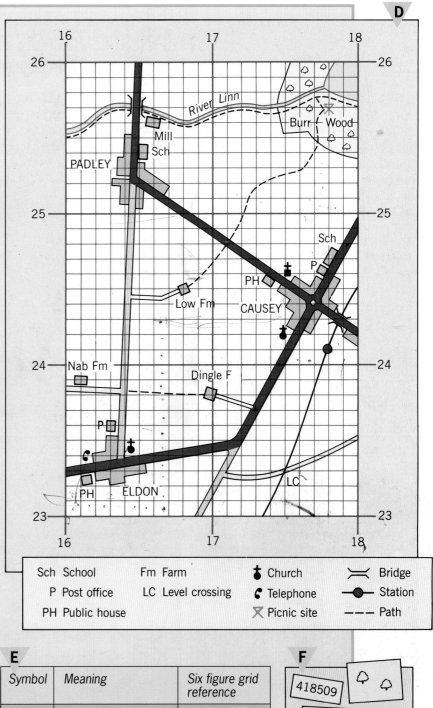

D

Sch	School	Fm	Farm	♟	Church	⧓	Bridge
P	Post office	LC	Level crossing	☏	Telephone	●—	Station
PH	Public house			✕	Picnic site	---	Path

E

Symbol	Meaning	Six figure grid reference
●—		465523
⌣		488505
	Church with tower	
	Camp/caravan site	

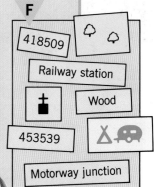

F

418509

Railway station

♟

Wood

453539

⛺

Motorway junction

440534

Summary

Six figure grid references can be used to give the exact position of a place on a map.

How is height shown on a map?

The land around us is seldom flat. There are nearly always differences in height and differences in slope. Sometimes slopes may be gentle and at other times steep. There may be hills, mountains and valleys or areas that are quite level. The word **relief** is used by geographers to describe the shape of the land.

Map makers have to find ways of showing relief and height. How they do this is shown on the next four pages.

Look at sketch **A**. How can height on the island be shown on a flat piece of paper? Height is usually measured from sea level in metres. This can then be shown on a map in three different ways. These are by using **spot heights, layer colouring** and **contours**.

A

B

Spot heights

These give the exact height of a point on the map. They are shown as a black dot and each one has a number next to it. The number gives the height in metres. A **triangulation pillar** is also used to show height. These are drawn as a dot inside a blue triangle on the map.

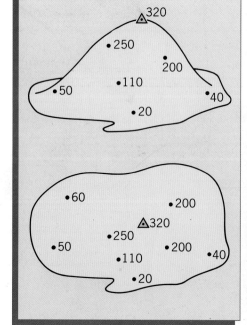

C

Layer colouring

This can also be called **layer shading**. Areas of different heights are shown by bands of different colours. Brown is usually used for high ground, and green for low ground. There always needs to be a key. Layer colouring is used in atlases to show height.

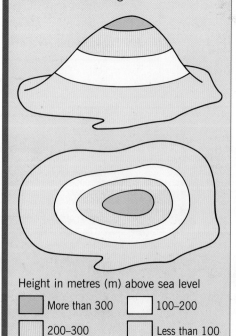

Height in metres (m) above sea level

▨ More than 300	☐ 100–200
☐ 200–300	☐ Less than 100

D

Contours

Contours are lines drawn on a map. They join places which have the same height. They are usually coloured brown. Most contours have their height marked on them but you may have to trace your finger along the line to find it. Sometimes you will have to go to the contour above or below to get the height. Heights are given in metres.

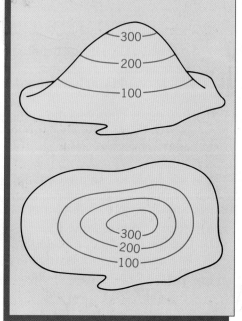

Activities

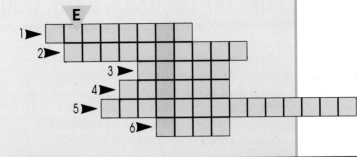

1 a) Copy out and complete crossword **E** using the clues below.

b) When you have finished, give the meaning of the downword in the orange squares.

Clues

1 Lines that join places of the same height.
2 Height at one place.
3 This can be gentle or steep.
4 Measured from sea level.
5 Colouring to show height.
6 A level area with no slope.

2 Look at map **F** of England and Wales. The map uses layer colouring to show height. The letters mark land at different heights.

a) Which letters mark lowland areas under 100 metres?

b) Which letters mark land between 100 and 300 metres?

c) Which letters mark land above 300 metres?

3 Use map **F** to answer these questions.

a) The highest mountain in England is Scafell Pike and the highest mountain in Wales is Snowdon. What colour are they shaded?

b) The Pennines are an area of high land in the centre of northern England. How high are they?

c) The Cotswolds and Chilterns are hills in the south of England. What height are they?

d) What height is the area where you live?

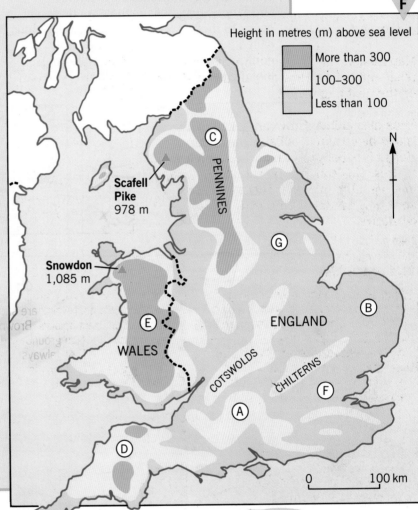

Height in metres (m) above sea level

More than 300
100–300
Less than 100

N

Scafell Pike 978 m

PENNINES

Snowdon 1,085 m

ENGLAND

WALES

COTSWOLDS

CHILTERNS

0   100 km

EXTRAS

Look at the Ordnance Survey map on page 109.

1 Give the heights above sea level of the following:
   a) the contours in grid squares 4852 and 4450
   b) the spot heights in grid squares 4151 and 4754
   c) the triangulation pillar in grid square 4051.

2 Look at Rowley's Hill in grid square 4249. Draw the pattern of contours and the triangulation pillar. Write in any heights that are given.

Summary

There are three main methods of showing height on maps. These are spot heights, layer colouring and contours.

How do contours show height and relief?

Lines on a map that join places of the same height are called **contours**. Contours show the height of the land and what shape it is. The shape of the land is called **relief**. The difference in height between contours is chosen by the map maker. On most Ordnance Survey maps they are drawn at every 10 metres. This difference in height is called the **contour interval**. Several contours together make up a pattern. By looking carefully at these patterns you can work out how steep the slopes are and what shape the land is.

Contour lines are drawn on maps by map makers; you cannot see them on the ground. In diagram **A** the contours have been drawn on the main sketch. You will see that they make up different patterns. An important thing to remember is that . . . *the closer the contour lines are together, the steeper the slope will be.*

A

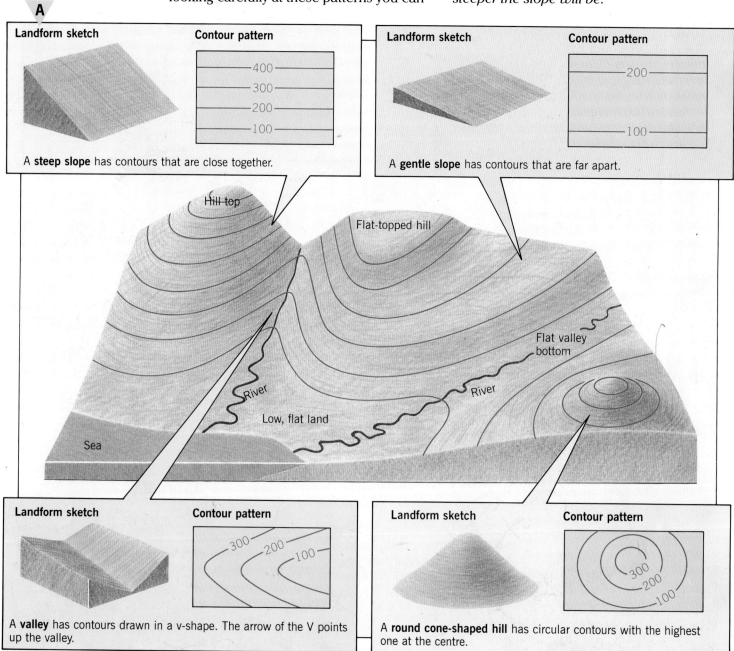

Landform sketch | **Contour pattern**
— 400 —
— 300 —
— 200 —
— 100 —

A **steep slope** has contours that are close together.

Landform sketch | **Contour pattern**
— 200 —
— 100 —

A **gentle slope** has contours that are far apart.

Hill top
Flat-topped hill
Flat valley bottom
River
River
Low, flat land
Sea

Landform sketch | **Contour pattern**
300 — 200 — 100

A **valley** has contours drawn in a v-shape. The arrow of the V points up the valley.

Landform sketch | **Contour pattern**
300 — 200 — 100

A **round cone-shaped hill** has circular contours with the highest one at the centre.

Activities

1 From map **B** give the heights of the following places. Choose your answers from those in the brackets.
   a) The highest point is (22, 48, 52, 40, 60) metres.
   b) Place **E** is (8, 42, 30, 20, 16) metres.
   c) Place **B** is (30, 20, 26, 46, 34) metres.
   d) Place **A** is (15, 10, 34, 6, 21) metres.
   e) Place **D** is (28, 10, 12, 22, 8) metres.

2 Look at map **B** and say if the following statements are TRUE or FALSE.
   a) **E** and **F** are at the same height.
   b) **D** is higher than **F**.
   c) **B** is higher than **E** but lower than **C**.
   d) **A** is the lowest place marked with a letter.
   e) **D** to **C** is steeper than **A** to **B**.

3 The photos in **C** show some landscape features.
   a) Draw a simple contour pattern for each of the photos.
   b) Write a description of the feature next to each of your drawings.

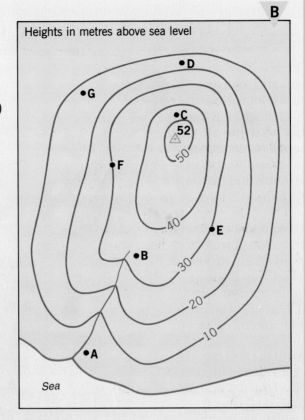

B

Heights in metres above sea level

•D
•G
•C
52
50
•F
40
•E
30
•B
20
10
•A

Sea

C

EXTRA

Look at the six areas circled on map **D**. Match the letters to each of the following:
1 A gentle slope
2 A steep slope
3 A hill top
4 A flat valley floor
5 A valley with a stream
6 A valley without a stream.

D

D
F
A
C
20
40
B
60
80
100
E

Height in metres above sea level

Summary

Contour lines are a good way of showing height and relief on a map. Contours that are close together show steep slopes. Contours that are far apart show gentle slopes.

How can we describe routes?

Maps show what things look like from above. They have a lot of information on them. You can use this to describe where places are and work out what may be seen there. Maps are also useful for describing routes between places. These two pages show how to describe routes and places from Ordnance Survey maps.

Paul lives in Foxton. He writes letters to a friend called Chris. Part of one of his letters is shown in **A**. It describes where he lives and a walk he often takes. Map **B** shows the area, which is near Cambridge. See if you can recognise the things Paul talks about. Can you follow the route?

Paul's description was good. He first described the area in general and then mentioned both the **physical** and **human** features. These are labelled on map **B**. When describing a place or a route, there is no need to try to include everything, but you must be very accurate.

A

Rose Cottage
Foxton

Dear Chris,
I live in the small village of Foxton. It has a church, Post Office and Public House. The area around here is open countryside and mainly flat. I often take my dog for a walk across to Newton which is 2½ km away. We follow a path across fields until we reach a stream and some trees. Rowley's Hill is to the north. It is 50m high and gently sloping. We then follow the stream and trees until the path becomes a narrow road. The road passes a church and some gardens belonging to the Manor before it reaches Newton. My walk usually ends at the Post Office.

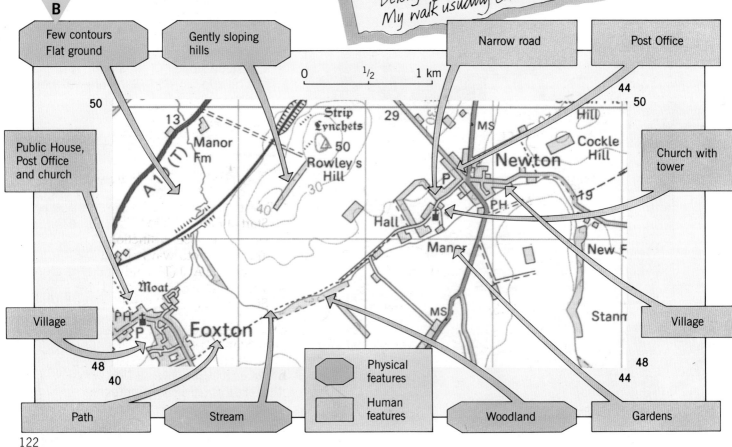

122

Activities

1 Look carefully at photo **C**.
  a) List **three** physical features and **three** human features on it.
  b) Write a brief description of the town.

2 Look at Newton on map **B**.
  a) List the physical features and human features in and around the village.
  b) Imagine that you own a cottage in the village that you want to rent out. Write a brief description of the village and surrounding area to advertise the cottage.

3 Map **D** shows part of the area covered by the OS map on page 125.
  a) Imagine that you have arrived by train at Great Shelford station and have to go to Hauxton post office. Some of the features that you will pass on your route are shown in diagram **E**. Write out features in the order you would pass them. Begin with the station.
  b) What is the distance from the station to the post office? Give your answer in kilometres.

C

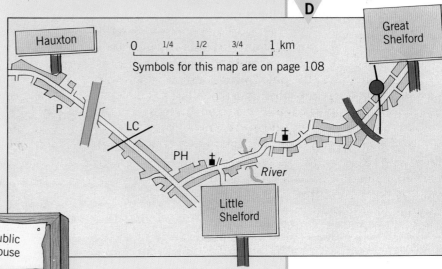

D

Hauxton

Great Shelford

0   1/4   1/2   3/4   1 km

Symbols for this map are on page 108

P

LC

PH

River

Little Shelford

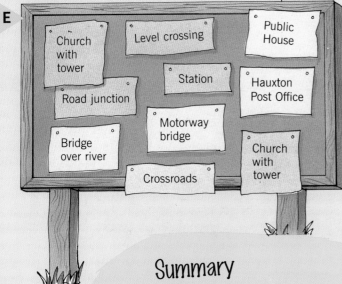

E

Church with tower

Level crossing

Public House

Station

Road junction

Hauxton Post Office

Motorway bridge

Bridge over river

Church with tower

Crossroads

E X T R A S

Use the Ordnance Survey map on page 125 for these questions.

1 Follow this route:
  Start at Barton (4055).
  Go on the A603 to junction 12.
  Travel by motorway to junction 11.
  Follow the A10 (T) south-west for 2 km.
  Turn south-east down the B1368.
  Stop after 2.5 km. Where are you?

2 a) Describe the route you would follow by road from Haslingfield (4052) to Grantchester.
  b) Describe the village of Grantchester and the surrounding area. Mention both physical and human features.

Summary

Maps can be used to describe routes and places. Accuracy is very important when describing things.

ROADS AND PATHS   Not necessarily rights of way

Service area   M 11   Elevated
Junction number   12   En Viaduc uberhoht

Motorway (dual carriageway)

Motorway under construction

Unfenced   Footbridge
A 11(T)

Trunk road

Dual carriageway
A 130

Main road

Main road under construction

B 1050

Secondary road

A 855   B 885

Narrow road with passing places

Bridge

Road generally more than 4 m wide

Road generally less than 4 m wide

Other road, drive or track

Path

Gradient : 1 in 5 and steeper
            1 in 7 to 1 in 5

Gates   Road tunnel

Ferry P   Ferry V

Ferry (passenger)   Ferry (vehicle)

PUBLIC RIGHTS OF WAY

(Not applicable to Scotland)

··················   Footpath

– – – – – – – –   Bridleway

–·–·–·–·–·–   Road used as a public path

–+–+–+–+–+–   Byway open to all traffic

Public rights of way indicated by these symbols have been derived from Definitive Maps as amended by later enactments or instruments held by Ordnance Survey on 1st May 1986 and are shown subject to the limitations imposed by the scale of mapping. Later information may be obtained from the appropriate County or London Borough Council

The representation on this map of any other road, track or path is no evidence of the existence of a right of way

Extent of available information

Danger Area   MOD Ranges in the area. Danger! Observe warning notices

TOURIST INFORMATION

i   Information centre   ▲   Youth hostel

P   Parking   Selected places of tourist interest

✕   Picnic site   ℓ ℓ   Telephone, public/motoring organisation

☼   Viewpoint   ⌐   Golf course or links

Å   Camp site   □   Public convenience (in rural areas)
                              PC

⌂   Caravan site

RAILWAYS

Track multiple or single

Track narrow gauge

Bridges Footbridge

Tunnel

Viaduct

Freight line, siding or tramway

Station   a   b   (a) principal
               (b) closed to passengers

Level crossing

LC

Embankment

Cutting

KILOMETRE   1   0   1
STATUTE MILE   1   0   1

WATER FEATURES

Marsh or salting   Cliff   High water mark
Towpath   Lock   Slopes   Flat rock   Low water mark
Aqueduct   Canal   Ford   Lighthouse (in use)
Weir   Normal tidal limit   Sand   Dunes   Beacon
Lake   Bridge   Footbridge   Mud   Lighthouse (disused)   Shingle

============   Canal (dry)

GENERAL FEATURES

Electricity transmission line (with pylons spaced conventionally)   Quarry

> – –> – –>   Pipe line (arrow indicates direction of flow)   Spoil heap, refuse tip or dump

ruin   Buildings   Radio or TV mast

Public buildings (selected)   ∎   Church   with tower
                                           or
Bus or coach station   ∎   Chapel   with spire

                       +   without tower or spire

Coniferous wood   ○   Chimney or tower

Non-coniferous wood   ⊘   Glasshouse

Mixed wood   +   Graticule intersection at 5' intervals

Orchard   H   Heliport

Park or ornamental grounds   △   Triangulation pillar

                       ✗   Windmill with or without sails

                       ⊥   Windpump

BOUNDARIES

–·–+–·+–   National   –·–·–·–·–   County, Region or Islands Area

–o–o–o–o–   London Borough   –+–+–+–+–   District

National Park or Forest Park

NT   National Trust   NT   always open

                              NT   opening restricted

FC   Forestry Commission   Pedestrians only observe local signs

ABBREVIATIONS

P   Post office   CH   Clubhouse
PH   Public house   PC   Public convenience (in rural areas)
MS   Milestone   TH   Town Hall, Guildhall or equivalent
MP   Milepost   CG   Coastguard

ANTIQUITIES

VILLA   Roman   ⚔   Battlefield (with date)   + Position of antiquity which cannot be drawn to scale
Castle   Non-Roman   ✲   Tumulus

▥   Ancient Monuments and Historic Buildings in the care of the Secretaries of State for the Environment, for Scotland and for Wales and that are open to the public

The revision date of archaeological information varies over the sheet

HEIGHTS

Contours are at 10 metres vertical interval
50

·144   Heights are to the nearest metre above mean sea level

ROCK FEATURES

outcrop
cliff
scree

Heights shown close to a triangulation pillar refer to the station height at ground level and not necessarily to the summit.

Coton
High Cross
Wimpole Way
Harcamlow Way
Laundry Fm
Whitwell Fm
Danger Area
IRE DISTRICT
Haggis Fm
Range
Danger Area
Cemy
Grantchester
Barton
Bird's Fm
Moat
Trumpington Hall
River Cam
River Fm
Clay A 1134
CAMBRIDGE DISTRICT
CAMBRIDGE
Coldham's Common
Cambridge Airp
Romsey Town Hospl
Govt Offices
Coll
Nature Reserve
Cherry Hinton
Netherhall Fm
Trumpington
Addenbrooke's Hospl
Anstey Hall
Weir
Byron's Pool
Cantelupe Fm
Spring Hall Fm
Travelling Telescope
Nine Wells
White Hill
Clarke's Hill
Caius
Nature Reserve
Stone Hill
Earthwork Moat
Middlefield
Fox Hill
Little Hill
Great Shelford
Haslingfield
Money Hill
Tumulus
Rectory Fm
Hauxton
Harston
Little Shelford
Stapleford
Bury Fm
River Granta
Dernford Fm
Obelisk
Harston Hill
Sainsfoins Fm
Rectory Fm
Works
Clunch Pit Hill
Cockle Hill
Strip Lynchets
Manor Fm
Rowley's Hill
Newton
Paper Mill
Hall
Manor
New Fm
Wells Fm
Stanmoor Hall
Foxton
Whittlesford

Reproduced from the 1994 Ordnance Survey map of Cambridge by permission of the Controller of HMSO © Crown Copyright.

Glossary

A	Accessibility	How easy a place is to get to. *70*
	Anticyclone	A weather system with high pressure at its centre. *26*
	Aspect	The direction which a slope or house faces. *20*
	Atmosphere	The air around the earth. *4*
B	Beaufort scale	A scale for measuring wind speed using things like smoke and trees. *18*
	By-pass	A road built around a busy area to avoid traffic jams. *78, 86*
C	Cargo	Goods which are carried by land, sea or air. *76*
	Central business district (CBD)	The middle of a town or city where most shops and offices are found. *62*
	Climate	The average weather conditions of a place. *5, 22, 95*
	Communications	The ways in which people, goods and ideas move from one place to another. *7*
	Condensation	The process by which water vapour changes to liquid water when cooled. *24, 34, 35*
	Congestion	Overcrowding on roads causing traffic jams. *71, 82, 83, 85*
	Contour	A line drawn on a map to join places at the same height about sea level. *118, 120*
	Contour interval	The difference in height between contours on a map. *120*
	Convectional rain	Rain that is produced when warm air rises. *24, 78*
	Corner shop	Small shops in inner city areas selling things which people need every day. *68, 69*
D	Deforestation	The cutting down or burning of trees to clear large areas of land. *39*
	Depression	A weather system with low pressure at its centre. *28*
	Direction	Shown on a map by the points of the compass. *19, 108*
	Dispersed settlement	Several farms or buildings spread out over a wide area. *56, 57*
	Drought	A long spell of dry weather. *95*
E	Economic geography	Is about industry, jobs, earning a living and wealth. *7, 100*
	Embankment	A raised river bank to prevent flooding. *46, 47*
	Environment	The natural or physical surroundings where people, plants and animals live. *8, 81, 90*
	Ethnic groups	People with similar culture, background and way of life. *99*
	Evaporation	The process by which liquid water changes to water vapour when warmed. *36*
F	Fieldsketch	A labelled sketch drawn outside of the classroom. *11*
	Floods	The flow of water over an area that is unusually dry. It may be a river flowing over flat land beside it, or the sea covering a low-lying coastal area. *38–51, 95*
	Four figure grid reference	A group of four figures to help find a square on an Ordnance Survey Map. *114*
	Freight	Goods which are carried by land, sea or air. *76*
	Frontal rain	When warm air has to rise over cold air in a depression. *25, 28*
	Function	The main purpose of a town. Functions include market and industrial towns, ports and resorts. *52, 62*
G	Goods	Materials and products that are useful to people, e.g. food, furniture, clothes, etc. *53, 68, 74, 76*
	Green belt	A protected area of countryside around a city. New building is controlled to try to stop the spread of the city. *82, 86*
	Grid square	A square on a map representing an area on the ground. *114*
	Groundwater	Fresh water stored in rocks and the soil. It may pass slowly through the rocks and soil back to the sea. *34, 35*
H	Habitat	The home of plants and animals. *5*
	Hazards	A natural danger to people and their property. Hazards include earthquakes, gales, drought and floods. *5*

Height	How high or low a place is. Measured in meters or feet above sea level. *118*
Hierarchy	Putting settlements and shops into order based upon their size or the services they give to people. *66, 67*
High and low order	These are goods sold in a shop. High order goods cost a lot but are not bought very often, e.g. furniture. Low order goods cost a lot but are bought more often, e.g. food. *68*
Human geography	Where and how people live. *6, 7*

I

Immigrant	A person who arrives in a country with the intention of living there. *99*
Industrial town	Where people make things like steel, cars and textiles in factories. *52*
Inner city	An area of factories and old houses next to the city centre. *62, 64*
Irrigation	The artificial watering of the land in a dry climate. *46*
Isobar	A line on a map joining places with the same atmospheric pressure. *28*

K

Key	A list of signs and symbols on a map or diagram with an explanation of what they mean. *112*

L

Land use	Describes how the land in towns or the countryside is used. It includes housing, industry and farming. *62*
Landforms	Natural features formed by rivers, the sea, ice and volcanoes. *4, 5*
Landscape	The scenery or appearance of an area. It includes both physical and human features. *122*
Latitude	This says how far north or south a place is from the Equator. *12, 13*
Layer colouring	A method of showing height on a map by using colours. *118*
Linear settlement	Buildings spread out in a line along a main road, a railway or a river. *56, 57*
Longitude	This says how far east or west a place is from the Greenwich Meridian. *12, 13*

M

Map	A drawing which shows part of the earth's surface from directly above, on a reduced scale. *12, 108*
Market town	The original function of the town was a place where people, mainly farmers, could buy and sell goods. *52*
Meteorology	The study of the weather. *18*
Microclimate	The climate of a small area. *20*
Migration	The movement of people from one place to another to live or work. *7, 98*
Multicultural society	A society where people with different beliefs and traditions live and work together. *98*

N

National parks	Areas of scenic beauty which are protected so that people can enjoy open air recreation. *87*
Network	A pattern of routes that are linked together. *80*
Network density	The number of routes that are linked together in an area. *80*
North Atlantic Drift	A warm ocean current that brings mild conditions to the west of Britain in winter. *22*
Nucleated settlement	Buildings which are grouped closely together. *56, 57*

O

Ordnance Survey	The official government organisation responsible for producing maps in the UK. *108, 112, 122*

P

Pattern	How things like settlements and shops are spread out over an area of land. *56, 57, 62, 120*
Physical geography	Natural features and events on earth. It includes landforms and weather. *4, 94*
Place	An area of the earth's surface. It can vary in size from a desk in a classroom to a city or a continent. *10*
Plan	A detailed map of a small area. *108*
Points of the compass	A method of giving direction using north, south, east, west, etc. *108*
Pollution	Noise, dirt and other harmful substances produced by people and machines which spoil an area. *9, 81, 82, 85*
Population	The people who live in an area. *6*
Port	A place used by ships to load and unload people and goods. *53*

Precipitation	Water in any form which falls to earth. It includes rain, snow sleet and hail. *18, 24, 34, 35*
Pressure	Atmospheric pressure is the weight of air pressing down on the earth's surface. *24, 26*
Public transport	Transport provided to the public and available to everyone, e.g. buses, trains, etc. *71*

Q

Quality of life	How content people are with their lives and the environment in which they live. *7, 48*

R

Refugees	People who have been forced to move from an area where they live, and have been made homeless. *99*
Relief	The shape of the land surface and its height above sea level. *102, 104*
Relief rain	Rain caused by air being forced to rise over hills and mountains. *22, 79*
Reservoir	An artificial lake used to store water. *41, 46, 48*
Resort	A place where people go for holidays. *45*
Resources	Things which are useful to people. They may be natural like coal and iron ore, and of value like money and skilled workers. *8*
Ribbon developments	Settlements that have a long narrow shape. *56*
River basin	An area of land drained by a river and its tributaries. *30, 32*
River channel	Where a river flows. It has a bed and two banks. *30*
River mouth	The end of a river where it enters the sea or a lake. *30*
River source	Where a river begins. *30*
Rural	An area of land which is mainly countryside. *77*

S

Scale	The link between the distance on a map and its real distance on the ground. *94*
Scale line	A short line on a map which shows how far real distances are. *94*
Settlement	A place where people live. *7, 44, 56, 82*
Shopping malls	Shopping areas which are undercover and protected from the weather. *58*
Site	The actual place where a settlement first grew up. *46, 80*
Six figure grid reference	A group of six figures used to give an exact position on a map. *100*
Spot height	A point on a map with a number giving its height above sea level in metres. *102*
Stores	Part of the water cycle where water is held in reserve in the sea, on land or in the air. *36*
Suburbanised village	A village with many new buildings added to it. *50*
Suburbs	A zone of housing around the edge of a city. *53*
Surface water	Water which lies on top of, or flows over, the ground. *28, 37*
Symbols	A simple drawing or sign used to give information and save space on a map. *17, 96*

T

Temperature	A measure of how warm or cold it is. *16, 20*
Transfers	Part of the water cycle when water moves between the sea, the land and the air. *36, 39*
Transpiration	The process by which water from plants changes into water vapour. *37*
Triangulation pillar	A concrete pillar used by surveyors to find the exact height and position of a place. *102*
Tributary	A small river which flows into a bigger river. *30*

U

Urban	An area of land which is mainly covered in buildings. *6, 48, 76*
Urban model	The pattern of land used in a town. *52*
Urbanisation	The growing proportion of people living in urban areas. *34*

V

Visibility	The distance that can be seen. *17*

W

Water cycle	The never ending movement of water between the sea, the land and the air. *28, 36*
Water demand	Water needed by people for domestic, farming and industrial use. *38*
Water supply	Water provided for people for domestic, farming and industrial use. *38*
Water shed	The boundary between two river basins. *30*
Weather	The day to day condition of the atmosphere. It includes temperature, rainfall and wind. *16*
Weathering	Rock is broken up by nature to give soil. *5*